Handmade Houses and Feeders for Birds, Bees, and Butterflies

Handmade Houses and Feeders for Birds, Bees, and Butterflies

35 havens for wildlife in your garden

Michele McKee-Orsini

CICO BOOKS
LONDON NEW YORK

Published in 2022 by CICO Books
An imprint of Ryland Peters & Small Ltd

20–21 Jockey's Fields
London WC1R 4BW

341 E 116th St
New York, NY 10029

www.rylandpeters.com

10 9 8 7 6 5 4 3 2 1

The designs in this book have previously appeared in one of the following titles by Michele McKee-Orsini: *Handmade Birdhouses and Feeders* or *Handmade Bird, Bee, and Bat Houses*.

A CIP catalog record for this book is available from the Library of Congress and the British Library.

ISBN: 978 1 80065 110 4

Printed in China

Editor: Sarah Hoggett
Designers: Mark Latter, Blue Dragonfly Ltd; Alison Fenton; and Jerry Goldie
Photographers: Caroline Arber, James Gardiner, and Emma Mitchell
Stylists: Sophie Martell, Joanna Thornhill, and Jess Contomichalos
Illustrator: Stephen Dew

In-house editor: Jenny Dye
Art director: Sally Powell
Head of production: Patricia Harrington
Publishing manager: Penny Craig
Publisher: Cindy Richards

contents

introduction

I've been building birdhouses and studying cavity-nesting birds since 2009. I am a true nature lover and would always rather be walking through the woods looking for interesting-looking bugs to study or a bird to watch. If I come across a snake in my path, I will help it get out of the way of on-comers who may do more harm than good. Yep… I pick them up and move them to a safer location (rattlesnakes I leave alone). I've even been known to take those creepy spiders out of the house and put them in my garden. I feel they have a family somewhere and I don't need to squish them. Sometimes I find myself talking to them as if they could understand me. Hah! Oh boy.

All of nature is essential for survival on this earth but, with the growth of buildings and people, there are many insects and animals that are in danger of disappearing. That's not good. If we keep destroying nature, we soon will have nothing to enjoy. I never take for granted the passing of a butterfly or the scampering squirrel who can't decide which way to go. I stop and watch until it disappears into the distance. I stop and smell the flowers, while looking at how their beauty is formed.

I am so grateful to have the capability to create and keep creating things I love and hold dear to my heart. The best thing for me is to find an object that someone else thought was trash and make it into a piece of artwork. I get excited over the smallest things. I remember when someone brought me an old rusty spoon. I jumped for joy! It made me feel good to know that I was in their thoughts and they knew I would be ecstatic with such a simple find.

In this book, I've been able to bring together my twin passions for nature and for recycling, to create a range of projects that will help you encourage birds and other wildlife into your own garden. My first book covered only birdhouses and bird feeders; this time around, I've expanded it to include bees, butterflies, and ladybugs, all of which we humans need to attract into our gardens. I am in hopes that this book brings you as much joy as it did me in building the creations and writing it.

In the year 2020 my husband, Guido, suddenly became ill and quickly passed away. This changed my life dramatically as he was my everything. He was always my biggest supporter. When he would see my work his praise was always over the top. I miss him terribly. Since then I have a new love in my life and life is good.

"Thank you for being in
a part of a bird's life."
Michele McKee-Orsini

basic birdhouse

Many of the projects in this book follow this basic design. It is made from dog-ear fence board (see page 126), which is a raw untreated wood that you can usually buy from your local home improvement store. The materials list in each project will tell you exactly what you need to buy, including any items needed for decoration. The list below is for a basic birdhouse without any embellishment. For more on decorating your birdhouse, turn to page 138.

materials

One 6ft x 5½ in. x ½ in. (180 x 14 x 1.2 cm) dog-ear fence panel
Waterproof premium glue
Wood putty
80-grit sandpaper
1-in. (25-mm) finish nails or galvanized wire nails
1¼-in. (30-mm) exterior screws
Basic tool kit (see page 126)

cutting list

As with the materials lists, each project will tell you the exact dimensions to cut the pieces to, as well as any other information such as whether you need to cut the pieces at an angle or with a beveled edge (see page 129).

Front and back: 9 x 5½ in. (23 x 14 cm)—cut 2
Sides: 6 x 5½ in. (16 x 14 cm)—cut 2 (one side is for the door)
Bottom roof panel 1: 5½ x 5½ in. (14 x 14 cm)—cut 1, then cut in half to give two panels measuring 5½ x 2¾ in. (14 x 7 cm)
Bottom roof panel 2: 6 x 5½ in. (15 x 14 cm)—cut 1, then cut in half to give two panels measuring 6 x 2¾ in. (15 x 7 cm)
Top roof panel 1: 6 x 5½ in. (15 x 14 cm)—cut 1
Top roof panel 2: 6¾ x 5½ in. (17 x 14 cm)—cut 1
Floor: 4¼ x 5½ in. (11 x 14 cm)—cut 1

Constructing the birdhouse

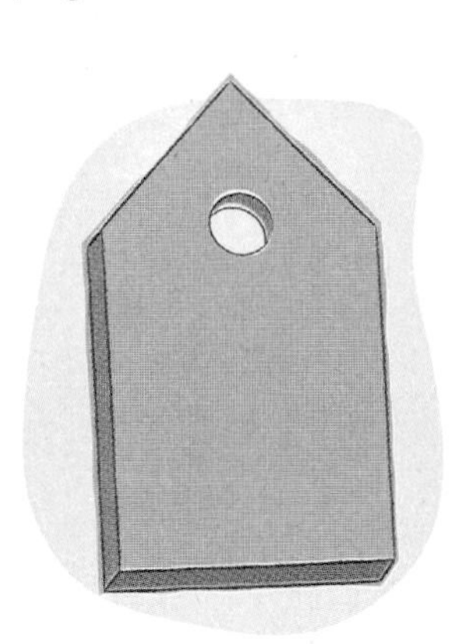

1 Cut the peaks of the front and back panels at a 45° angle. Place the front panel on a stable surface with a cutting board or scrap piece of wood underneath and drill the entrance hole, using a hole saw and the appropriate size of bit (see page 130). (See page 140 for the desired height and hole diameter for specific bird species.)

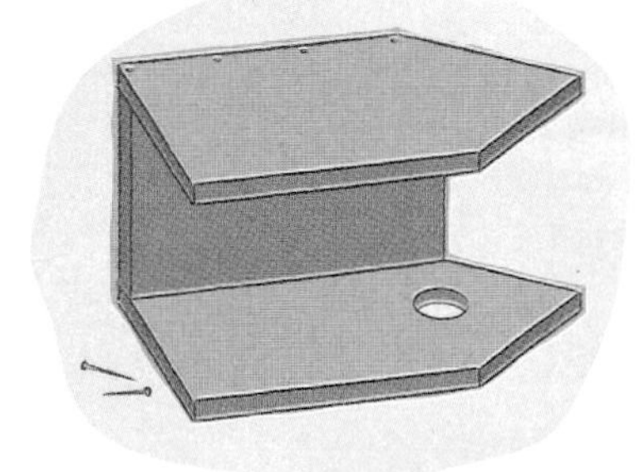

2 Glue and nail one side panel to the inside edges of the front and back panels. Leave the other side open for the door.

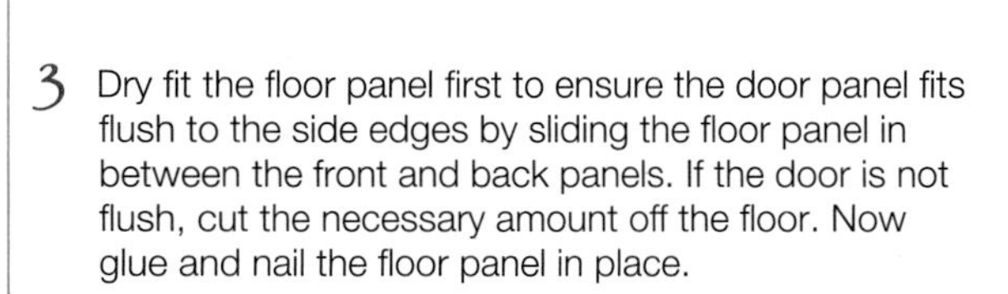

3 Dry fit the floor panel first to ensure the door panel fits flush to the side edges by sliding the floor panel in between the front and back panels. If the door is not flush, cut the necessary amount off the floor. Now glue and nail the floor panel in place.

4 Place the door panel between the front and back panels. Use a speed square to line up where the pilot holes will be placed on the front and back edges: the screws need to be ¾ in. (2 cm) down and directly across from each other in order for the door to swing open properly. Using a ¼-in. (6-mm) bit, drill a pilot hole (see page 131) through the front and back panels into the door below the point at which the roof slopes upward. Insert a 1¼-in. (30-mm) exterior screw at each point to act as hinges for the door.

5 Line up a bottom roof panel 1 with the roof peak flush with the inside edge of the panel, then position bottom roof panel 2 overlapping it, as shown. The roof pieces will overhang slightly at the front and sides, protecting the entrance hole during rainy weather. Glue and nail in place.

6 Glue and nail the other set of bottom roof panels flush with the outside edge of the back panel for a wall mount or overhanging the back panel by ¼ in. (6 mm) if you are going to hang the birdhouse from its roof.

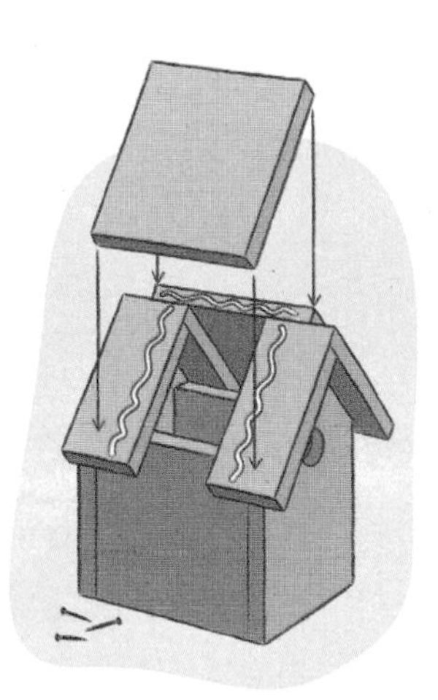

7 Attach the top roof panels in the same way, centering them on the bottom roof, which creates vent ducts.

8 Fill all holes with paintable wood filler putty, then sand the birdhouse lightly with 80-grit sandpaper to knock off any splinters and smooth the putty flush with the wood.

9 The dowels are optional and are merely for decorative purposes. Drill a hole, using a 5⁄16-in. (8-mm) drill bit. Cut a 5⁄16-in. (8-mm) wooden dowel rod to 1¼ in. (30 mm) in length. Hammer the dowel rod into the hole until only ¼ in. (6 mm) protrudes on the outside of the birdhouse.

10 Prepare for painting by brushing off any excess dust and placing tape or wadded paper in the hole. This keeps the inside of the hole and the interior of the birdhouse free of paint, as it is important for the birds that the inside is clean, natural wood.

simple roof

This roof design is even simpler and quicker to construct than the one shown above. Simply cut the roof pieces to the required size, then follow the steps below.

1 The project instructions will tell you whether the roof pieces should be flush with the back or front panels of the birdhouse or whether they should overhang. Glue and nail the roof pieces in place, so that they touch along the center ridge. This will leave a V-shaped gap on the roof ridge.

2 Cut a piece of 1 x 1-in. (2.5 x 2.5-cm) square dowel to the length required. Glue and nail it into the V-shaped gap, angling the nails slightly so that they go through both the dowel and the roof piece. (Again, the individual project instructions will tell you whether the dowel is to be positioned flush with the back/front of the birdhouse or to overhang.)

chapter 1

birdhouses

In this chapter you'll find a variety of fun and stylish ideas, such as a rustic seaside birdhouse decorated with shells and driftwood, and a triple-decker home ideal for small cavity-nesting birds. From simple to novelty designs, these projects will provide ample room for birds to nest and raise their young.

Attract small cavity-nesting birds to your garden with this rustic but simple birdhouse. A wire attached to the roof makes it easy to hang. Small pieces of driftwood are used on the front façade and for the door knob, with moss tucked in between the cracks.

simple birdhouse

materials

One 6 ft x 5½ in. x ½ in. (180 x 14 x 1.2 cm) dog-ear fence panel
Waterproof premium glue
1-in. (25-mm) finish nails or galvanized wire nails
Two 1¼-in. (30-mm) exterior screws
Three 1¼-in. (30-mm) exterior screws
1¾-in. (4.5-cm) length of round dowel, 5⁄16 in. (8 mm) in diameter
10½-in. (26.5-cm) length of 1 x 1-in. (2.5 x 2.5-cm) square dowel
26-in. (66-cm) length of heavy wire
Paintable wood-filler putty
80-grit sandpaper
Gray wood-primer paint
Yellow oil-based exterior spray paint
Dark brown exterior craft paint
Small pieces of driftwood
Moss
Water-based exterior varnish
Basic tool kit (see page 126)

finished size

Approx. 10½ x 6½ x 5½ in. (26.5 x 16.5 x 14 cm)

interior dimensions

Floor area: 4¼ x 5½ in. (10.75 x 14 cm)
Cavity depth: 8 in. (20 cm)
Entrance hole to floor: 6 in. (15 cm)
Entrance hole: 1½ in. (40 mm) in diameter

cutting list

Front and back: 9¼ x 5½ in. (23.5 x 14 cm)—cut 2
Sides: 6½ x 5½ in. (16.5 x 14 cm)—cut 2 (one side is reserved for the door)
Roof: 8¾ x 5½ in. (22 x 14 cm)—cut 2
Floor: 4¼ x 5½ in. (11 x 14 cm)—cut 1

1 Referring to the Basic Birdhouse on pages 8–9, cut and shape the birdhouse pieces from dog-ear fence board; do not bevel the roof panels. Cut a 1½-in. (40-mm) entrance hole, 6½ in. (16.5 cm) from the bottom of the front panel and centered on the width.

2 Assemble the body of the birdhouse, following steps 2–4 of the Basic Birdhouse on page 8. Attach the roof, following the instructions for the Simple Roof on page 9, with the square dowel overlapping each end of the roof by 1 in. (2.5 cm). Using a 5⁄16-in. (8-mm) bit, drill a hole in the front panel, 1 in. (2.5 cm) below the entrance hole, and insert a length of round dowel for the perch (see step 9 on page 9).

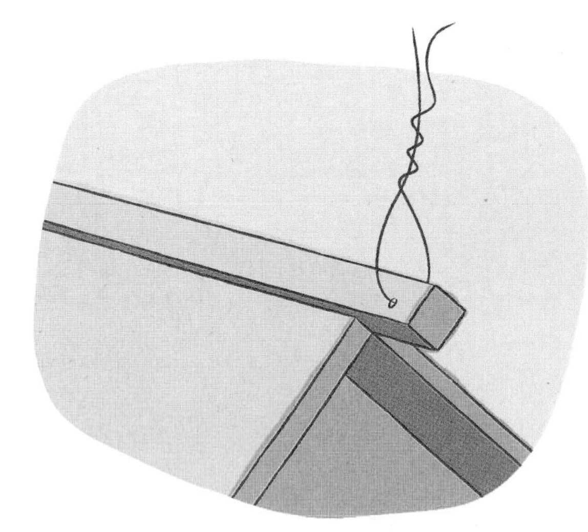

3 Drill a ⅛-in. (3-mm) hole through each overlapping end of the square dowel on the roof. Thread the end of the wire through and twist several times to secure. Make a hanging loop in the center of the wire by twisting the wire around several times.

4 Prepare the birdhouse for painting (see step 10 on page 9). Allowing the paint to dry between coats, prime the birdhouse all over (including the base) with gray paint, then paint the body of the box yellow. Lightly sand the body of the birdhouse so that some of the gray shows through. Dilute dark brown exterior paint, 1 part paint to 2 parts water, and paint the roof. Sand the edges of the birdhouse and roof to give a worn look.

5 Using a ⅛-in. (3-mm) bit, drill pilot holes for screws in the driftwood sticks for the front panel. Countersink the screw heads (see page 131), then screw in place, taking care not to overtighten. Apply a little glue and press on moss to hide the screw heads. Now apply a little glue to the side of the driftwood and tuck moss under and around the edge. Pre-drill a pilot hole through the back of the door into a small piece of driftwood for the door knob. Screw the door knob in place.

6 Plug the entrance hole with wadded-up paper or tape, then varnish the exterior of the birdhouse (see page 127).

Tufted Titmouse

This brightly colored birdhouse will accommodate many types of small cavity-nesting birds. The amount of decoration you add is entirely up to you, but the hardware (door strike plate, metal tacks, and bent fork) gives it a retro, industrial feel that is right on trend.

retro birdhouse

materials

- Two 6 ft x 5½ in. x ⅝ in. (180 x 14 x 1.5 cm) cedar dog-ear fence panels
- Waterproof premium glue
- Wood putty
- 80-grit sandpaper
- 1-in. (25-mm) finish nails or galvanized wire nails
- Two 1¼-in. (30-mm) exterior screws
- Cardstock for masking
- Black, blue, and red exterior spray paint
- Water spray bottle
- Door strike plate
- Two 1¼-in. (30-mm) decorative bronze screws
- Three round dowels, 1½ in. (4 cm) long and 5⁄16 in. (8 mm) in diameter
- Fork
- Three #18 (¾-in./20-mm) escutcheon pins
- Ceramic door knob
- Nut and washer
- Six black carpet tacks
- Basic tool kit (see page 126)

finished size

Approx. 12 x 7¼ x 10 in. (30 x 18.5 x 25 cm)

interior dimensions

Floor area: 4¼ x 5½ in. (10.8 x 14 cm)
Cavity depth: 10 in. (25 cm)
Entrance hole to floor: 6 in. (15 cm)
Entrance hole: 1¼ in. (30 mm) in diameter

cutting list

Front and back: 10 x 5½ in. (25 x 14 cm)—cut 2, cutting one end of each at a 15° angle
Left side: 9½ x 5½ in. (24 x 14 cm)—cut 1
Right side (door): 8¼ x 5½ in. (21 x 14 cm)—cut 1
Floor: 4¼ x 5½ in. (10.8 x 14 cm)—cut 1
Platform: 10 x 5½ in. (25 x 14 cm)—cut 1
Stairs: Find the center of the fence board, then cut off the two corners at 45° to give two right-angle triangles
Door façade: 2 x 2¼ in. (5 x 5.7 cm)—cut 1
Bottom roof: 7 x 5½ in. (18 x 14 cm)—cut 1, with one short end beveled at 15°, then cut in half lengthwise
Top roof: 7 x 5½ in. (18 x 14 cm)—cut 1, with one short end beveled at 15°
Side roof: 9½ x 3 in. (24 x 7.5 cm)—cut 1

1. Cut a 1¼-in. (30-mm) entrance hole in the front panel 1½ in. (4 cm) in from the right-hand side and 6½ in. (16.5 cm) up from the bottom (see page 130). Assemble the body of the birdhouse, following steps 1–4 on page 8.

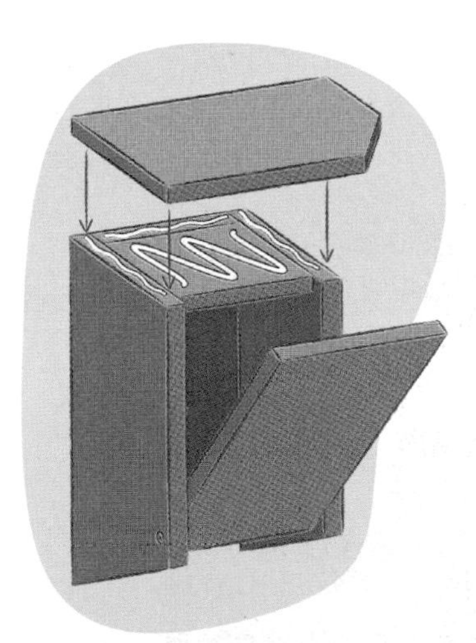

2. Find the center of one short side of the platform, then cut off the corner from this point at 45°. Reserve the cut-off triangle for the steps (see step 3). With the door open, apply glue to the bottom of the birdhouse and then position the platform flush with the back and sides, as shown. Nail in place.

A store-bought bird table makes a good base, but don't scatter bird seeds on it. Never put food close to or in the birdhouse.

3 Position the first step on the platform, with one short side flush to the angled platform edge, and mark with a pencil where the next two steps will go; the last one needs to be flush against the front of the birdhouse. When you're happy with the positioning, glue and nail the steps together from the back so that the nail holes won't be seen.

4 Paint the steps and door façade red (or your chosen color) and let dry. Set aside.

5 Attach the bottom and main roof panels to the long sloping edges, following steps 5–7 on page 9. Attach the side roof, lining it up with the top roof edge.

6 Carefully mask off the roof and bottom platform with cardstock and spray the body of the birdhouse black. While the paint is still wet, fill a spray bottle with water. Working one side at a time, spray gently with water, then with blue paint, and then spray once more with water.

7 Paint the roof and platform red and let dry.

This is a very messy process! Place the birdhouse on a plank of wood over a trash can to catch any paint drips and overspray.

8 Glue and nail the stairs in place, nailing them from the underside so that you won't have nail holes on the front. Glue the door façade on top of the stairs. Drill pilot holes (see page 131) ⅛ in. (3 mm) from the top and bottom of the door and ¼ in. (6 mm) from the left-hand side, then insert decorative screws. Attach the door strike plate in the same way.

9 Cut three dowels 1½ in. (4 cm) long. Drill three evenly spaced ⁵⁄₁₆ in. (8-mm) holes on the front, 1 in. (2.5 cm) from the entrance hole. Hammer the dowels in place, leaving approx. ½ in. (1 cm) protruding. Paint the dowels black and let dry.

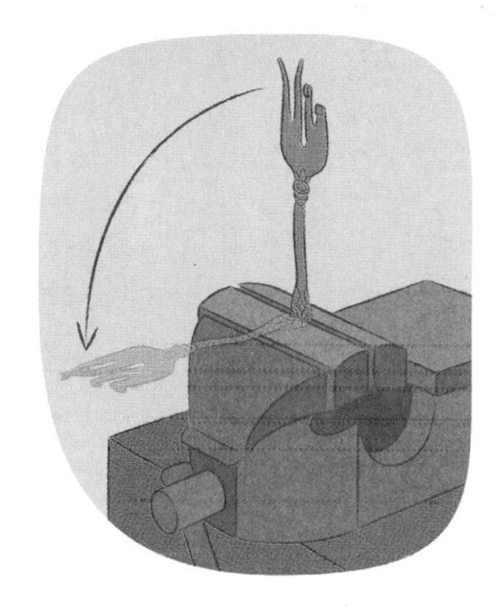

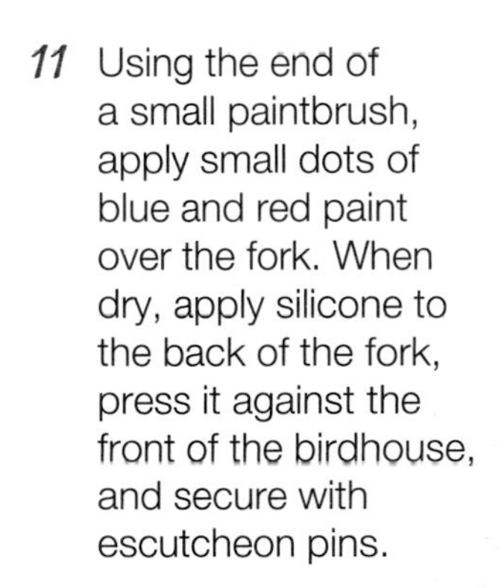

10 Find a fork that has an opening at the bottom so that you can secure it with a nail. Using needle-nose pliers, bend the tines of the fork as shown. Place the fork in a vise and bend the middle of the handle into a half-moon shape. Bend the end of the handle forward until it's flat. Paint black on all sides and let dry.

11 Using the end of a small paintbrush, apply small dots of blue and red paint over the fork. When dry, apply silicone to the back of the fork, press it against the front of the birdhouse, and secure with escutcheon pins.

12 Using a ⅜-in. (10-mm) bit, drill a hole in the door 1½ in. (4 cm) up from the bottom. Attach the door knob, securing it on the back with a nut and washer.

13 Drill ¹⁄₁₆-in. (1.5-mm) pilot holes in the roof edges and stairs, as shown in the photo, then gently hammer in black carpet tacks.

14 Place on a hollow metal post 6 ft (2 m) above the ground, in concrete (see page 135), or on a store-bought bird table.

The vibrant green color and three-tiered roof will make this birdhouse a talking point in any garden. Hanging it from a piece of driftwood is a suitably rustic-looking touch and ties in visually with the decorative elements on the front of the birdhouse.

triple-roof hanging birdhouse

materials

One 6 ft x 5½ in. x ½ in. (180 x 14 x 1.2 cm) dog-ear fence panel
Waterproof premium glue
1-in. (25-mm) finish nails or galvanized wire nails
1¼-in. (30-mm) exterior screws
1¾-in. (4.5-cm) length of round dowel, 5⁄16 in. (8 mm) in diameter
Paintable wood-filler putty
80-grit sandpaper
Brown wood primer paint
Green oil-based exterior spray paint
Five driftwood pieces, approx. 3–7 in. (7.5–18 cm) long
1-in. (25-mm) EMT two-hole pipe clamp (clip)
Two ¾-in. (20-mm) wood screws
Two 7-in. (18-cm) length of heavy wire
Exterior craft paints (optional)
Water-based exterior varnish
Basic tool kit (see page 126)

finished size

Approx. 12 x 6¾ x 5½ in. (30 x 17 x 14 cm)

interior dimensions

Floor area: 4¼ x 5½ in. (10.75 x 14 cm)
Cavity depth: 8 in. (20 cm)
Entrance hole to floor: 6 in. (15 cm)
Entrance hole: 1¼ in. (32 mm) in diameter

cutting list

Front and back: 9¼ x 5½ in. (23.5 x 14 cm)—cut 2
Sides: 6¼ x 5½ in. (16 x 14 cm)—cut 2 (one side is reserved for the door)
Bottom roof: 5 x 5½ in. (13 x 14 cm)—cut 2
Center roof: 5¾ x 5½ in. (14.5 x 14 cm)—cut 2
Third roof: 4 x 5¼ in. (10 x 13.5 cm)—cut 2
Floor: 4¼ x 5½ in. (11 x 14 cm)—cut 1

1 Referring to the Basic Birdhouse on page 8, cut and shape the birdhouse pieces from dog-ear fence board. Cut a 1¼-in. (32-mm) entrance hole, 6½ in. (16.5 cm) from the bottom of the front panel and centered on the width.

2 Assemble the birdhouse, following steps 2–8 of the Basic Birdhouse on pages 8–9, then glue and nail the third roof on top of the center roof.

3 Drill a hole and insert a length of round dowel for the perch (see step 9 on page 9).

4 Prepare the birdhouse for painting (see step 10 on page 9). Allowing the paint to dry between coats, prime the birdhouse all over (including the base) with brown paint, then paint the body of the box with green paint. Lightly sand the body of the birdhouse so that some of the brown shows through. Dilute dark brown exterior craft paint, 1 part paint to 2 parts water, and paint the roof, leaving the front trim green. Leave to dry. Sand the edges of the roof to give a worn look.

5 Attach a 1-in. (25-mm) EMT two-hole pipe clamp (clip) to the apex of the roof for a hanging loop (see page 132).

6 Take a 7-in. (18-cm) length of driftwood stick and drill a small hole at each end for the hanging wire. Thread one wire through one hole and then through the loop of the pipe clamp (clip), and twist both ends of the wire together with pliers to secure. Form another wire loop at the other end of the driftwood for hanging the birdhouse.

7 Remove the door. Using a ⅛-in. (3-mm) bit, drill pilot holes (see page 131) through from the back, then attach one driftwood piece to the door to act as a knob and others to the front for decoration, using 1¼-in. (30-mm) screws. Replace the door. If you wish, paint flowers on the front panel, using exterior craft paints.

8 Plug the entrance hole with wadded-up paper or tape, then varnish the exterior of the birdhouse (see page 127).

Black Capped Chickadee

The perfect patio birdhouse to accommodate beloved songbirds such as the chickadee or tits. These birds are friendly and not afraid to nest once they feel safe. Not only will your little birds enjoy this condo, it will also dress up the patio.

cottage birdhouse condo

materials

- Two 6 ft x 5½ in. x ⅝ in. (180 x 14 x 1.5 cm) cedar dog-ear fence panels
- Waterproof premium glue
- Wood putty
- 80-grit sandpaper
- 1-in. (25-mm) finish nails or galvanized wire nails
- Eight 1¼-in. (30-mm) exterior screws
- Peach, soft pink, white, and gray exterior spray paint
- Cup and saucer
- Tile saw
- Weatherproof 30-minute clear silicone
- Styrofoam or florist's foam
- Green moss
- Silk flowers
- Nine #18 (¾-in./20-mm) escutcheon pins
- Driftwood pieces
- Three metal dragonfly adornments
- Black and metallic green craft paints (optional)
- Two glass knobs
- Two nuts and washers
- Basic tool kit (see page 126)

finished size

Approx. 14¼ x 14 x 9½ in. (36 x 35.5 x 24 cm)

interior dimensions

Main birdhouse

Floor area: 4¼ x 5½ in. (10.8 x 14 cm)
Cavity depth: 12 in. (30 cm)
Entrance hole to floor: 10 in. (25 cm)
Entrance hole: 1¼ in. (30 mm) in diameter

Side birdhouse

Floor area: 4¼ x 5½ in. (10.8 x 14 cm)
Cavity depth: 9¾ in. (24.7 cm)
Entrance hole to floor: 6¼ in. (16 cm)
Entrance hole: 1¼ in. (30 mm) in diameter

cutting list

Main birdhouse

Front and back: 12 x 5½ in. (30 x 14 cm)—cut 2
Right side panel: 9½ x 5½ in. (24 x 14 cm)—cut 1
Left side panel (door): 8½ x 5½ in. (21.5 x 14 cm)—cut 1
Floor: 4¼ x 5½ in. (10.8 x 14 cm)—cut 1
Platform: 9 x 5½ in. (23 x 14 cm)—cut 1

Side birdhouse

Front and back: 9¾ x 5½ in. (24.7 x 14 cm)—cut 2
Door: 4 x 5½ in. (10 x 14 cm)—cut 1
Left side panel: 8½ x 5½ in. (21.5 x 14 cm)—cut 1
Floor: 4¼ x 5½ in. (10.8 x 14 cm)—cut 1
Platform: 9 x 5½ in. (23 x 14 cm)—cut 1

Roof

Bottom roof panels (long side) 1A and 1B: 12 x 5½ in. (30 x 14 cm)—cut 1, then cut in half lengthwise to give two pieces measuring 12 x 2¾ in. (30 x 7 cm)
Bottom roof panels (short side) 2A and 2B: 6 x 5½ in. (15 x 14 cm)—cut 1, then cut in half lengthwise to give two pieces measuring 6 x 2¾ in. (15 x 7 cm)
Main roof panel (long side) 3: 12⅝ x 5½ in. (32 x 14 cm)—cut 1
Main roof panel (short side) 4: 6⅝ x 5½ in. (17 x 14 cm)—cut 1

1 First, cut the peaks of the front and back panels of the main birdhouse at an angle. Working from right to left, measure 3 in. (7.5 cm) along the top edge of the front panel and mark this point. From your mark, cut the top edge at a 45° angle on your chop saw. Now measure 2 in. (5 cm) down the left-hand side of the panel, draw a line from here to the peak, and cut. Cut the back panel in the same way, reserving all the cut-off triangles.

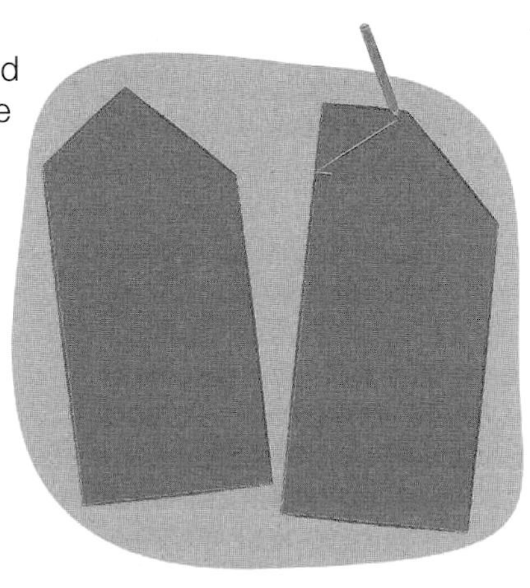

2 Paint the smaller cut-off triangles from step 1 peach, let dry, and set aside. These will form the planters on the front of the birdhouse in step 14.

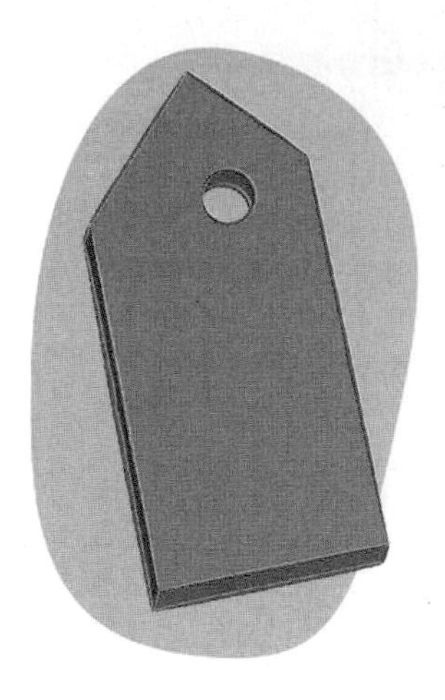

3 Cut a 1¼-in. (30-mm) entrance hole in the front panel, 3 in. (7.5 cm) down from the peak and 2 in. (5 cm) in from the shorter side (see page 130).

4 Assemble the body of the birdhouse, following steps 1–4 on page 8.

5 Cut off about ¾ in. (2 cm) from the two front corners of the platform at a 45° angle. With the birdhouse door open so that you don't accidentally nail it shut, glue and nail the platform to the base of the birdhouse, aligning the straight short edge with the back.

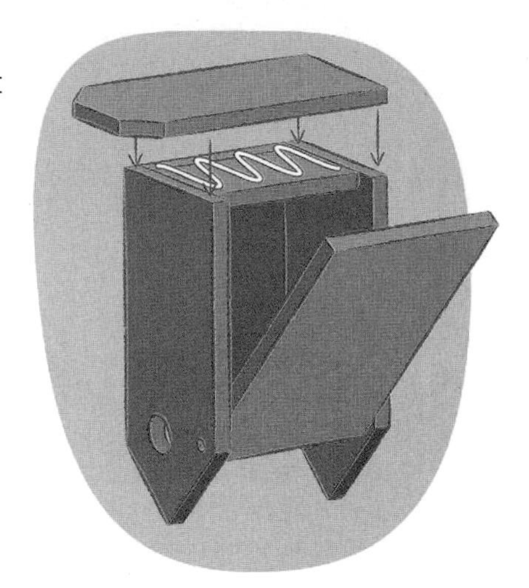

6 Now make the smaller side birdhouse. Cut the peaks of the front and back pieces at a 45° angle. Cut a 1¼-in. (30-mm) entrance hole in the front panel, 3¾ in. (9.5 cm) down from the peak and 1¾ in. (4.5 cm) in from the longest side (see page 130). Place the front and back pieces against the main birdhouse to ensure that the points match up with the long sloping side.

7 Assemble the side birdhouse, following steps 1–3 on page 8. (Don't attach the door at this stage.)

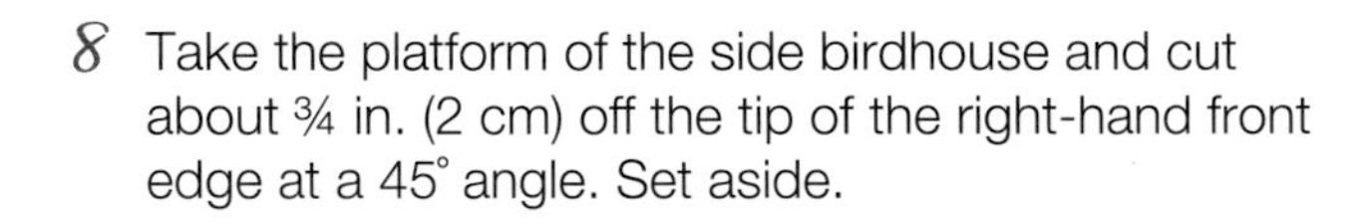

8 Take the platform of the side birdhouse and cut about ¾ in. (2 cm) off the tip of the right-hand front edge at a 45° angle. Set aside.

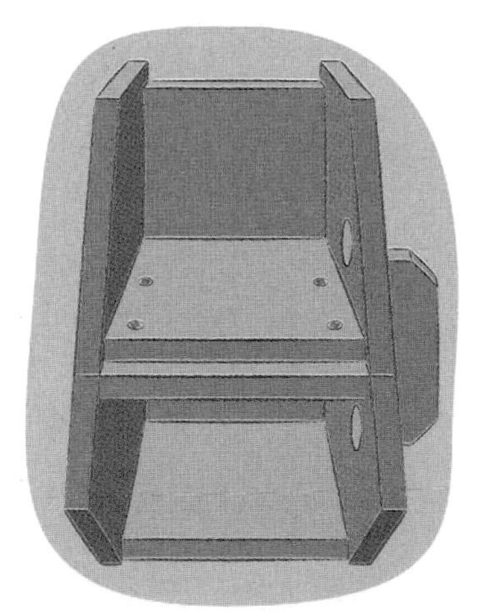

9 Place both birdhouses together to check that the sloping sides align. Apply glue to the side of the smaller birdhouse. Drill four pilot holes (see page 131) for exterior screws in the side panel of the smaller birdhouse, making sure both birdhouses are snug to each other. Drive in the screws.

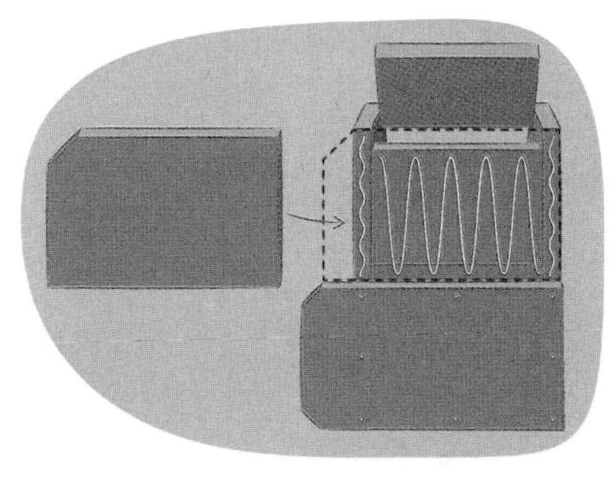

10 Attach the door of the smaller birdhouse, following step 4 on page 8. Open the door of the smaller birdhouse. Apply glue to the bottom of the smaller birdhouse, then position the platform flush with the back edge and snug against the platform of the main birdhouse, with the angled cut on the side nearest the door. Nail in place.

11 Attach the bottom and main roof panels, one side at a time, following steps 5–7 on page 9.

12 Paint the birdhouse roof and platform soft pink, and the body white. Let dry. Drybrush gray paint over the entire birdhouse to create a slightly weathered look.

13 Using a tile saw, cut the cup in half. Stick it onto the saucer with clear silicone, wiping off any excess with a cotton swab, and let dry. Apply a good amount of silicone to the bottom of the saucer, then stick it onto the birdhouse front and let dry.

14 Cut the Styrofoam or florist's foam to size and glue it into the cup with silicone, below the rim. Let dry. Glue moss around the rim of the cup, under the saucer, and on the platform, then insert silk flowers into the foam. To make the planters, take the painted triangles from step 2, drill a couple of 1/16-in. (1.5-mm) pilot holes, then attach to the birdhouse with escutcheon pins. Glue moss on top. Finally, add driftwood sticks and the painted dragonfly embellishments.

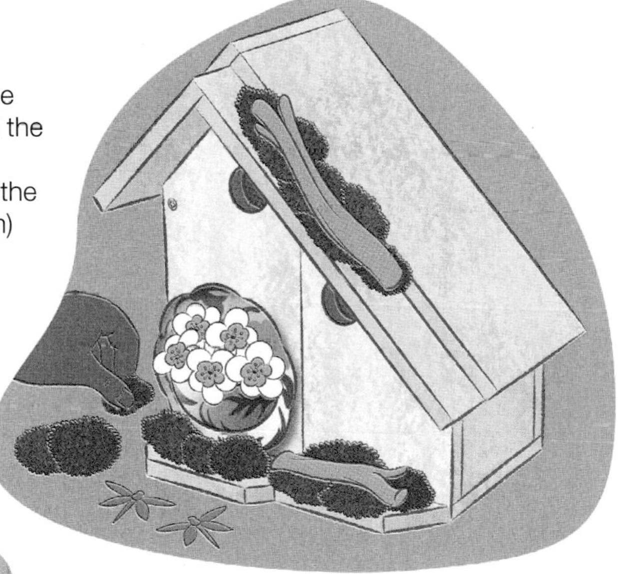

You can buy plain, gilt-colored dragonfly embellishments in the jewelry section of your local craft store. If you want to make them look really realistic, use craft paints to paint the whole piece soft pink, then paint the tips of the wings black, and the rest of the wings and the body metallic green.

15 Using a 3/16-in. (5-mm) bit, drill a hole 1½ in. (4 cm) above the bottom edge of each clean-out door, and insert a door knob, securing them on the inside with a nut and washer.

Traveling to the beach is a fond memory of mine, and picking up all sorts of shells and pieces of driftwood gave me the inspiration for a nautical-themed birdhouse.

seaside birdhouse

materials

Two 6 ft x 5½ in. x ½ in. (180 x 14 x 1.2 cm) dog-ear fence boards
Waterproof premium glue
1-in. (25-mm) finish nails or galvanized wire nails
Two 1¼-in. (30-mm) exterior screws
Two 2-in. (50-mm) exterior screws
Three lengths of 5/16-in (8-mm) round dowel, 1, 1½, and 1¾ in. (2.5, 4 and 4.5 cm) long respectively
Three 5½ x ½-in. (14 x 1.2-cm) strips of dog-ear fence panel
18¼ x 3½ in. (46.5 x 9 cm) dog-ear fence board
Paintable wood-filler putty
80-grit sandpaper
Dark brown and ivory oil-based exterior spray paints
Brown exterior craft paint
Water-based exterior varnish
Seaside-themed embellishments of your choice—for example, star fish, shells, rope ½ in. (12 mm) in diameter
3 skeleton keys (optional)
Weatherproof silicone glue and glue gun and exterior glue sticks
Basic tool kit (see page 126)

finished size

Approx. 14 x 5½ x 6⅜ in. (35.5 x 14 x 16 cm)
Back panel mount: 18½ x 5½ in. (50 x 14 cm)

interior dimensions

Floor area: 4¼ x 5½ in. (11 x 14 cm)
Cavity depth: 11½ in. (29 cm)
Entrance hole to floor: 10 in. (25 cm)
Entrance hole: 1¼ in. (32 mm) in diameter

cutting list

Front and back: 12¼ x 5½ in. (31 x 14 cm)—cut 2
Sides: 9½ x 5½ in. (24 x 14 cm)—cut 2 (one side is reserved for the door)
Bottom roof: 5 x 5½ in. (20 x 14 cm)—cut 2
Top roof: 5¾ x 5½ in. (14.5 x 14 cm)—cut 2
Floor: 4¼ x 5½ in. (11 x 14 cm)—cut 1
Back panel mount: 18-½ x 5½ in. (50 x 14 cm)—cut 1 from dog-ear end of fence board

1 Referring to the Basic Birdhouse on page 8, cut and shape the birdhouse pieces from dog-ear fence board. Cut a 1¼-in. (32-mm) entrance hole, 10 in. (25 cm) from the bottom of the front panel and centered on the width. Then, assemble the birdhouse and prepare it for painting, following steps 2–8 of the Basic Birdhouse on pages 8–9.

2 Using a 5/16-in. (8-mm) bit, drill three holes below the entrance hole for perches, spacing them randomly. Tap the 1-in. (2.5-cm) length of dowel into the left-hand hole, the 1½-in. (4-cm) length into the center hole, and the 1¾-in. (4.5-cm) length into the right-hand hole, leaving each one sticking out slightly further than the previous one.

3 Cut three 5½ x ½-in. (14.5 x 1.2-cm) pieces from fence panel for the front ledges. Lay the first piece across the roof from edge to edge, mark on the back, then cut at a 45° angle on each side. Glue and nail in place. Glue and nail the second piece 5 in. (13 cm) above the base of the front panel. Cut the third piece to approx. 3 in. (7.5 cm) long and at a 45° angle at one end to make the last ledge. Glue and nail it 1½ in. (4 cm) from the bottom.

4 Using a ⅛-in. (3-mm) bit, drill two pilot holes in the dog-ear end of the mount board for 2-in. (5-cm) exterior screws. Using a ⅛-in. (3-mm) bit, drill pilot holes through the back of the birdhouse, then drill in 1¼-in. (30-mm) exterior screws to attach the birdhouse to the back mount board.

5 Paint the birdhouse and back mount panel dark brown and leave to dry. Next, paint the top coat in ivory and leave to dry. Then paint the dowel perches and ledges dark brown and wipe with a rag while wet to get a translucent effect. Dilute 1 part brown exterior craft paint to 2 parts water and paint the roof to create a worn effect.

6 Using weatherproof silicone glue or a glue gun with exterior glue sticks, attach your chosen embellishments wherever you choose. If you're using rope or small keys, nail them in place for extra security.

7 Drill a hole through the door with the appropriate size of bit. Push the neck of the door knob through the hole and attach with a washer and a nut.

With its mossy green covering and pieces of driftwood and wood bark, this birdhouse is inspired by Mother Nature. Set it on a patio table or a shelf under a garden porch; alternatively, attach a metal pipe clamp (clip) to the roof as a hanger (see page 132).

moss-covered birdhouse

materials

One 6 ft x 5½ in. x ½ in. (180 x 14 x 1.2 cm) dog-ear fence panel
Waterproof premium glue
1-in. (25-mm) finish nails or galvanized wire nails
1¼-in. (30-mm) exterior screws
Paintable wood-filler putty
80-grit sandpaper
Green oil-based exterior spray paint
Self-adhesive moss sheet, available from craft stores
Loose, artificial moss
Driftwood sticks and small wood bark chips
Staples
Basic tool kit (see page 126)

finished size

Approx. 11 x 6⅜ x 5½ in. (28 x 16 x 14 cm)

interior dimensions

Floor area: 4¼ x 5½ in. (11 x 14 cm)
Cavity depth: 9 in. (23 cm)
Entrance hole to floor: 6 in. (15 cm)
Entrance hole: 1½ in. (40 mm) in diameter

cutting list

Front and back: 9¼ x 5½ in. (23.5 x 14 cm)—cut 2
Sides: 6¼ x 5½ in. (16 x 14 cm)—cut 2 (one side is reserved for the door)
Bottom roof: 5 x 5½ in. (13 x 14 cm)—cut 2
Top roof: 5¾ x 5½ in. (14.5 x 14 cm)—cut 2
Floor: 4¼ x 5½ in. (11 x 14 cm)—cut 1

1 Referring to the Basic Birdhouse on page 8, cut and shape the birdhouse pieces from dog-ear fence board. Cut a 1½-in. (40-mm) entrance hole, 6½ in. (16.5 cm) from the bottom of the front panel and centered on the width.

2 Assemble the birdhouse, following steps 2–8 of the Basic Birdhouse on pages 8–9. Prepare the birdhouse for painting (see step 10 on page 9), then paint green and leave to dry. Lightly sand.

3 Cut the self-adhesive moss sheet to size, peel off the backing, and apply to the birdhouse one panel at a time, making sure you don't cover the sides of the door. Use a utility knife to cut a slit in the moss over the front entrance hole, then push the moss through to the inside with your fingers. Glue loose moss into any gaps, then staple the moss around the edges of the birdhouse to fix it securely.

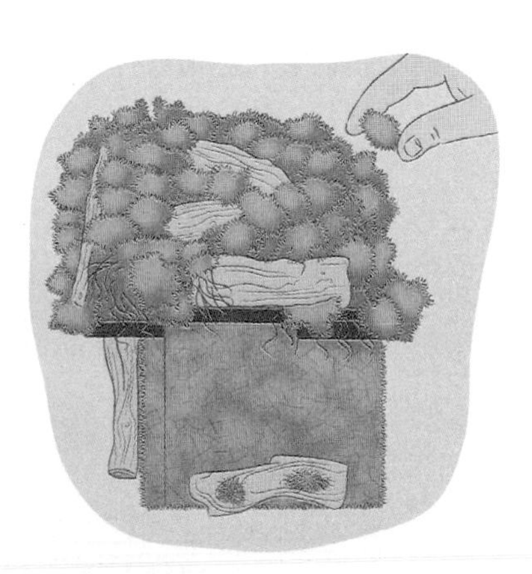

4 Pre-drill holes in driftwood sticks, then screw the sticks onto the roof and box. Remember to attach a piece of driftwood for the door knob. Apply a generous amount of glue to the roof and cover with loose moss and small pieces of bark, pressing the moss down firmly to ensure that it sticks.

The "weathered look" painted roof and decoration add a touch of natural-looking color and perfectly complement the raw wood of the birdhouse.

Invite owls into your garden with a cute home where they can nest and thrive. The owls will look over the property, keeping rodents out, leaving your flowers, fruits, and vegetables to be enjoyed by you alone.

Screech Owl

owl house

materials

Two 6 ft x 7½ in. x ¾ in. (180 x 19 x 2 cm) cedar dog-ear fence panels
Five pine stakes, 12 x 3 x 1 in. (30 x 7.5 x 2.5 cm)
Waterproof premium glue
Wood putty
80-grit sandpaper
1¼-in. (30-mm) finish nails or galvanized wire nails
Four decorative 1¼-in. (30-mm) screws
Yellow and green craft paint
Two 1¼-in. (30-mm) exterior screws
Approx. 3 in. (8 cm) 18-gauge (1-mm) rebar tie wire
Basic tool kit (see page 126)
Jigsaw
Wire cutters

finished size

Approx. 16 x 9 x 9 in. (40.5 x 23 x 23 cm)

interior dimensions

Floor area: 7½ x 7½ in. (19 x 19 cm)
Cavity depth: 16½ in. (42 cm)
Entrance hole to floor: 12½ in. (32 cm)
Entrance hole: 3 in. (7.5 cm) in diameter

cutting list

Floor: 7½ x 7½ in. (19 x 19 cm)—cut 1
Sides: 12⅝ x 7½ in. (31 x 19 cm)—cut 2, with one short end of each beveled at 45°
Support triangles: 3½ x 2½ x 2½ in. (9 x 6 x 6 cm)—cut 4 from scrap or the end of your fence board
Upper and lower back: 9 x 7½ in. (23 x 19 cm)—cut 2
Lower front: 9 x 7½ in. (23 x 19 cm)—cut 1
Upper front: 9 x 7¼ in. (23 x 18.5 cm)—cut 1
Roof panel 1: 10 x 7 in. (25 x 17.8 cm)—cut 1
Roof panel 2: 10 x 7½ in. (25 x 19 cm)—cut 1

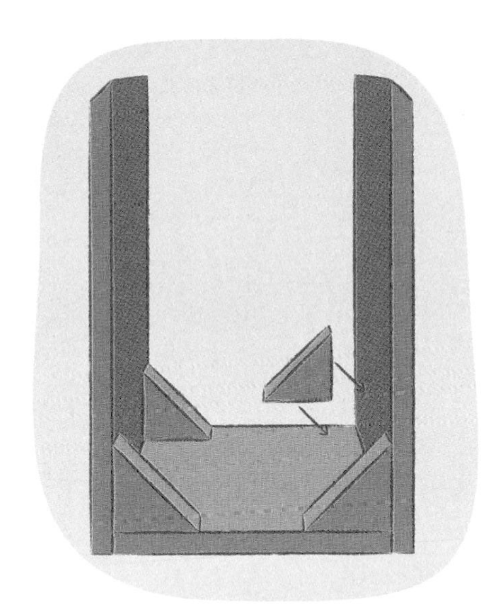

1 Glue and nail the sides to the outside edge of the floor, with highest points of the beveled edges on the inside. Glue and nail a support triangle to each internal corner to ensure that the bottom and sides are square.

2 Glue and nail the lower front and back pieces in place, making sure they line up and the edges are flush.

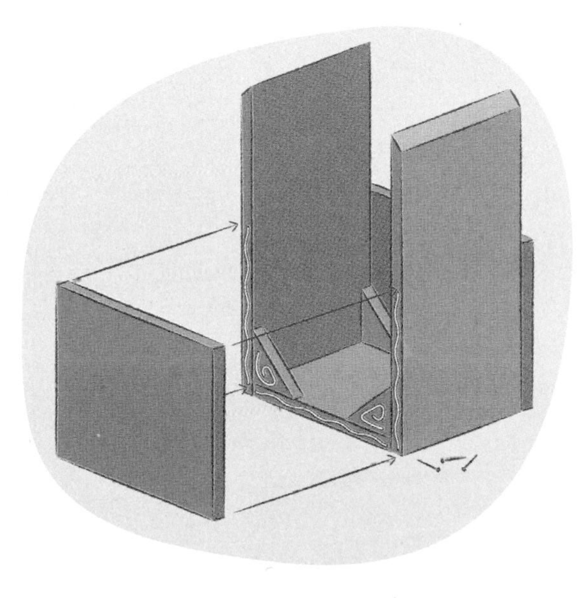

3 Place the upper front and back pieces in position and draw a line across where they touch the beveled side panels. Cut off the resulting right-angle triangles at these points and reserve for later. Apply glue to the side and bottom edges of the upper back panel, line it up so it's flush with the side edges, and nail in place.

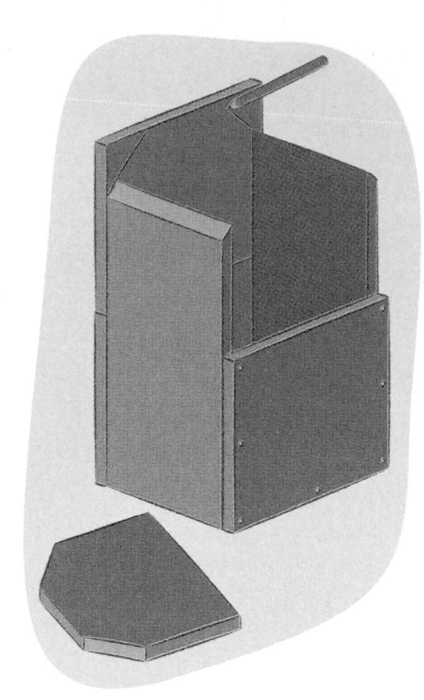

Dry fit the upper back before gluing it in place, to ensure it's flush with the beveled side edges where the roof will be attached.

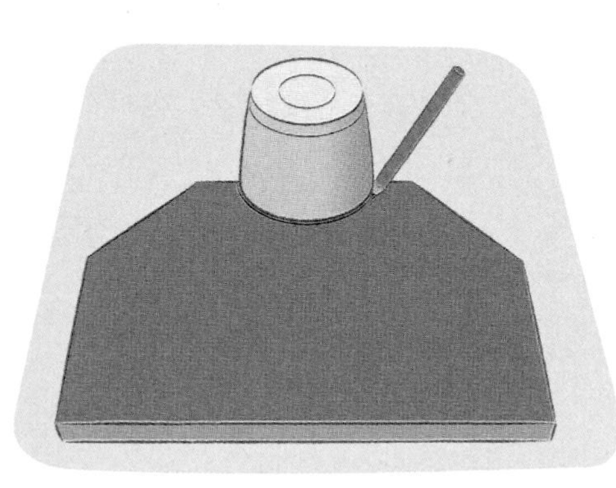

4 Find the center of the upper front panel between the angled edges. Using a lid or something round that measures 3 in. (7.5 cm) in diameter, mark out a half-moon. Cut out with a jigsaw.

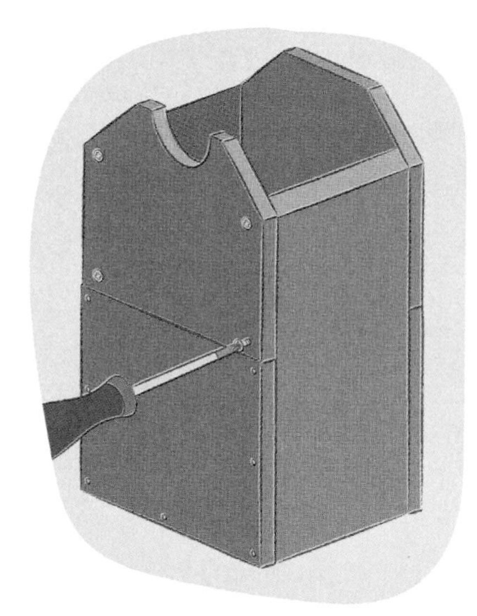

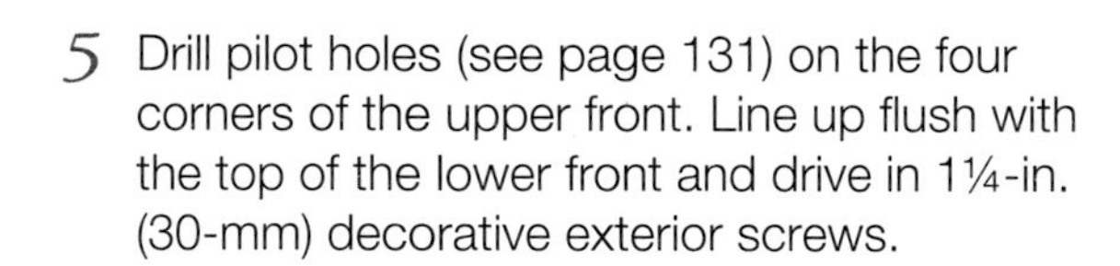

5 Drill pilot holes (see page 131) on the four corners of the upper front. Line up flush with the top of the lower front and drive in 1¼-in. (30-mm) decorative exterior screws.

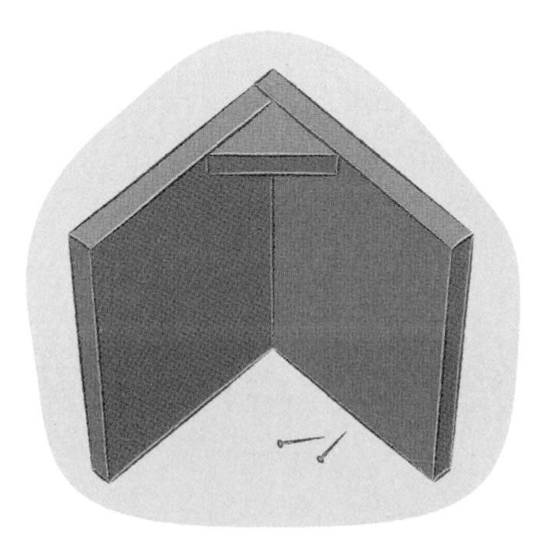

6 Assemble the roof (see page 9). Glue and nail one of the reserved right-angle triangles from step 3 to the back of the roof to square it and add strength.

7 Temporarily remove the upper front panel in order to attach the roof. Glue and nail the roof at an angle so that the nails do not protrude through the side panels. Re-attach the upper front panel.

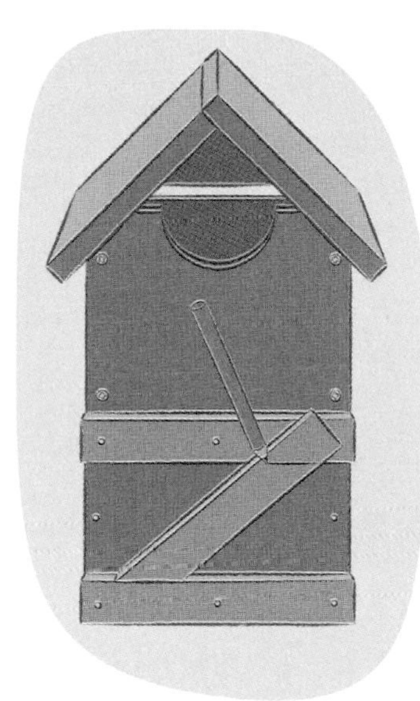

8 Cut five stakes to approximately 9 in. (23 cm) in length, reserving the pointed tips for decoration. Glue and nail one 9-in. (23-cm) stake ⅛ in. (3 mm) below the "seam" between the upper and lower front panels, and another across the bottom of the lower front. Place a third stake diagonally across them and mark where to cut at a 45° angle to complete the "Z." Glue and nail this last piece in place.

9 Center one of the reserved pointed tips about ½ in. (12 mm) below the hole, and glue and nail it in place. Add two more tips on either side, curving them in a semicircular shape.

10 Paint the roof, "Z," and three of the pointed tips dark green and the remaining tips yellow, using a drybrush technique to give a slightly uneven coverage. Let dry, then drybrush yellow very lightly over the dark green areas. Lightly sand all the painted areas to create a slightly worn texture.

11 Drill a pilot hole (see page 131) on each side of the back of the roof, about 2 in. (5 cm) below the apex. Attach screws and a wire to form a hanging loop (see page 132).

House Sparrow

Simple birdhouses joined together make a classic condo for birds to enjoy. I chose three different shades of enticing blue for my condo, but you could make each birdhouse a completely different color if you prefer. I used pieces of driftwood as embellishments, as I love using "found" objects as decoration and I think they give the birdhouses a lovely rustic feel—but the type and amount of decoration you add is entirely up to you.

birdhouse *trio*

materials

One 6 ft x 7½ in. x ½ in. (180 x 19 x 1.2 cm) dog-ear fence panel
Two 6 ft x 5½ in. x ½ in. (180 x 14 x 1.2 cm) dog-ear fence panels
Waterproof premium glue
1-in. (25-mm) finish nails or galvanized wire nails
Ten 1¼-in. (30-mm) exterior screws
Three 1¾-in. (4.5-cm) lengths of 5/16-in. (8-mm) round dowel
Driftwood sticks
Paintable wood-filler putty
80-grit sandpaper
Oil-based exterior spray paints in three shades of blue (or your chosen color)
Dark brown and dark green exterior craft paints
Water-based exterior varnish
Basic tool kit (see page 126)

finished size

Approx. 8½ x 19½ x 5½ in. (21.5 x 49.5 x 14 cm)

interior dimensions

Floor area: 4¼ x 5½ in. (10.75 x 14 cm)
Cavity depth: 8 in. (20 cm)
Entrance hole to floor: 6 in. (15 cm)
Entrance hole: 1½ in. (4 cm) in diameter

cutting list

Front panel (door): 8 x 5½ in. (20 x 14 cm)—cut 1 for each box
Back panel: 8½ x 5½ in. (21.5 x 14 cm)—cut 1 for each box
Side panels: 8¾ x 5½ in. (22.3 x 14 cm)—cut 2 for each box
Floor: 4¼ x 5½ in. (11 x 14 cm)—cut 1 for each box
Side roof panels: 6½ x 7½ in. (16.5 x 19 cm)—cut 2
Center roof panel: 9 x 7½ in. (23 x 19 cm)—cut 1

1 Referring to the Basic Birdhouse on page 8, cut the birdhouse pieces from dog-ear fence board. Cut the top edges of the side panels at a 15° angle. Bevel the top edges of the front and back panels at 15°. Cut a 1½-in. (4-cm) entrance hole, 1½ in. (4 cm) from the top of each front panel and centered on the width.

2 Following steps 2–4 of the Basic Birdhouse on page 8, assemble the bodies of the three birdhouses; the tallest side of the side panels should align with the back panel, while the front panel – the door – will overhang at the base by 1 in. (2.5 cm). On each front panel, drill a hole 1 in. (2.5 cm) below the entrance hole and insert a length of round dowel for the perch (see step 9 on page 9). Remove the front panels (the doors) of all three boxes, as this makes it easier to join them together.

3 On the box that will go in the middle of the condo, using a speed square, draw a straight line vertically down the center of each side. Now measure 3½ in. (9 cm) up the center line from the base and draw a line from here to the front edge. The other two boxes will attach at these points. Using a ⅛-in. (3-mm) bit, drill two pilot holes (see page 131) through the side panels of the outer and center boxes from the inside and insert 1¼-in. (30-mm) screws.

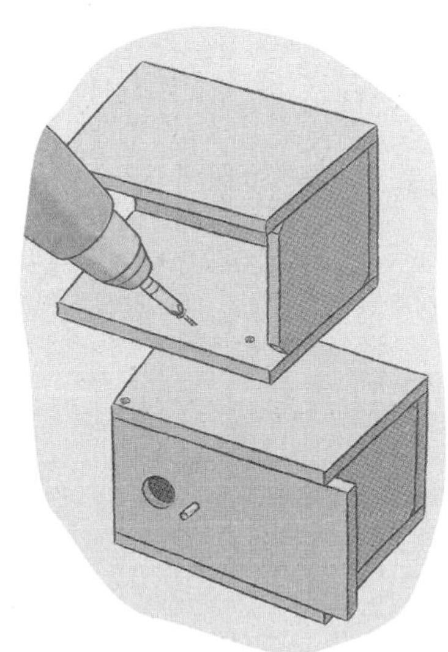

4 Glue and nail the roofs of the two outer boxes in place, flush with the sides of the center box and overhanging on the other three sides. Glue and nail the center roof panel to the middle box, overhanging by the same amount on all four sides. Re-attach the three door panels.

5 Prepare the birdhouse for painting (see step 10 on page 9). Paint the body of each birdhouse in a different shade of blue (or color of your choice). Paint the roofs and their edges dark brown.

6 While the brown paint is still wet, brush streaks of dark green across from back to front to create the impression of moss growing on the roof. Wipe off any excess paint if necessary, taking care to wipe in the same direction as your brushstrokes. Leave to dry, plug the entrance hole with wadded-up paper or tape, then varnish the exterior of the birdhouse (see page 127).

7 Lay the assembled birdhouse on your work surface with the front facing up. Place driftwood sticks on top to check the position, making sure that the doors will open once they are attached. Remove the doors one at a time. Drill pilot holes (see page 131) from the back, then screw the driftwood pieces in place. Attach sticks to the top of the middle birdhouse roof in the same way, then replace the doors.

8 Place the birdhouse on a patio or sturdy, level wall. Alternatively, you could mount it on a post (see page 135). Mounting birdhouses on galvanized steel poles is the best way to keep predators from climbing up in search of young nestlings or fledglings. You can find the poles and everything you need to attach the birdhouse to them in the fencing and electrical departments of your local home-improvement store.

It's always great to find a use for all those objects that you just couldn't bear to throw away. On this unusually shaped birdhouse, I used S-hooks, metal screws, composite shims, a screwdriver that came with a kit of some sort, and metal roof flashing. You could omit the embellishments and hang the birdhouse from a pipe clamp (clip) if you prefer. There is an opening on the bottom that the hub connector will cover for cleaning out.

pyramid birdhouse

materials

- One 6 ft x 5½ in. x ½ in. (180 x 14 x 1.2 cm) dog-ear fence panel
- Waterproof premium glue
- 7-in. (18-cm) length of 1 x 1-in. (2.5 x 2.5-cm) lumber (timber)
- 1-in. (25-mm) finish nails or galvanized wire nails
- Paintable wood-filler putty
- 80-grit sandpaper
- Purple oil-based exterior spray paint
- Water-based exterior varnish
- Eight 1½-in. (38-mm) S-hooks
- Nine ⅝-in. (15-mm) metal screws
- Three packs composite shims
- ¾-in. (20-mm) wire nails (for composite shim attachment)
- 5 x 7-in. (12.5 x 18-cm) roof flashing tile
- Ten ¾-in. (20-mm) wood screws (for metal roof)
- Metal screwdriver, key, or 1¾-in. (4.5-cm) length of 5⁄16-in. (8-mm) round dowel
- 1-in. (2.5-cm) HS hub connector (from electrical section of home store)
- Four 1¼-in. (30-mm) exterior screws
- Four ¼-in. (6-mm) nuts
- 1 x 6-in. (2.5 x 15-cm) galvanized nipple
- 6-ft (1.8-m) hollow galvanized fence pipe
- Two 4-foot (1.2-m) rebars (reinforcing bars), 5⁄16 in. (8 mm) in diameter
- Basic tool kit (see page 126)

finished size

7½ x 15 x 5½ in. (19 x 38 x 14 cm)

interior dimensions

Floor area: 4¼ x 10 in. (11 x 25 cm)
Cavity depth: 5 in. (12.5 cm)
Entrance hole to floor: 4 in. (10 cm)
Entrance hole: 1 in. (25 mm) in diameter

cutting list

Front and back: 7⅝ x 7⅝ x 10¾ in. (18.5 x 18.5 x 27.3 cm)—cut 2
Roof: 9 x 5½ in. (23 x 14 cm)—cut 2
Bottom panel: 12¼ x 5½ in. (31 x 14 cm)—cut 1

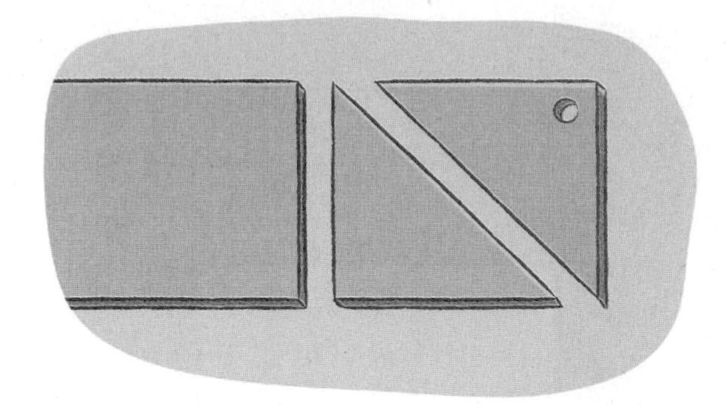

1 Cut two triangle shapes for the front and back panels, cutting across the straight end of a dog-ear fence panel at a 45° angle, as shown. Cut a 1-in. (2.5-cm) entrance hole in the front panel 1½ in. (4 cm) from the peak and centered on the width.

2 Glue and nail the two roof panels flush to the edge of the back panel on each side, then glue and nail the front panel flush to the edges of the roof panels on each side, leaving a V-shaped gap along the ridge of the roof.

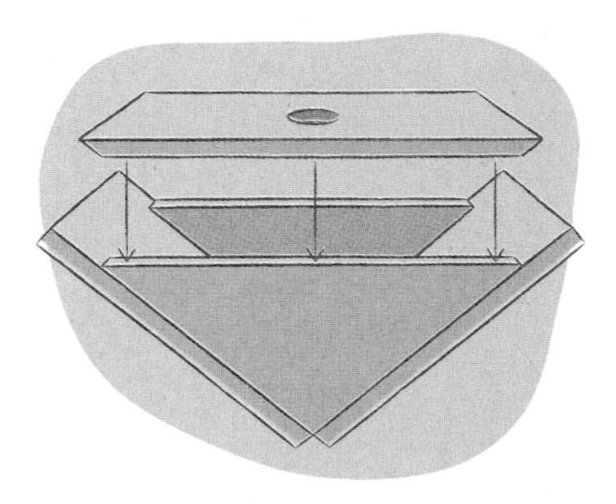

3 Bevel each short end of the bottom panel at 45°. Cut a 2½-in. (6.5-cm) hole in the center of the bottom panel for cleaning out the birdhouse when necessary. Glue and nail the bottom panel to the front, back, and roof sides.

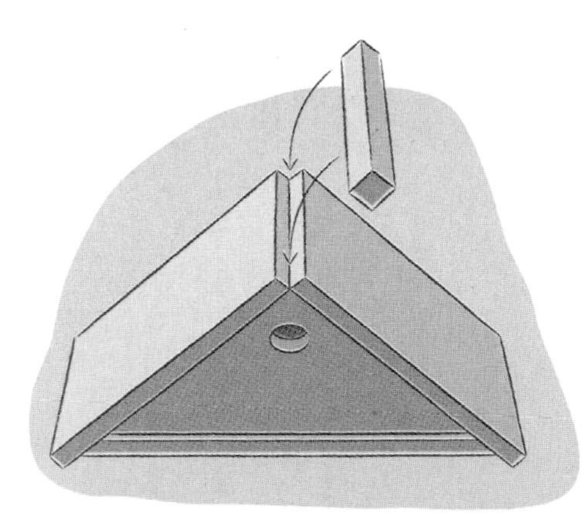

4 Glue and nail a 7-in. (18-cm) length of 1 x 1-in. (2.5 x 2.5-cm) square lumber to the V-shaped gap in the ridge of the roof, with the end flush with the back panel and the tip overhanging at the front. (Nail the piece at a slight angle to ensure proper attachment.)

5 Prepare the birdhouse for painting (see step 10 on page 9), then spray with purple oil-based exterior paint.

6 Center the hub connector hole over the clean-out hole in the base. Using a ⅛-in. (3-mm) bit, drill a pilot hole in the center of each hole of the hub connector. Drill in 1¼-in. (30-mm) exterior screws with ¼-in. (6-mm) nut on the neck of each screw to prevent sharp tip from penetrating through to the floor bottom. Screw the 6-in. (15-cm) length of nipple to the hub connector.

7 Measure the composite shims against the roof and cut them to size, using utility scissors. Each row will hold four shims, with a slight overhang at the sides and at the bottom. Starting in the center of the roof and working outward, glue and nail the first row in place, using ¾-in. (20-mm) nails. Attach three more rows in the same way, starting each row 1½ in. (4 cm) higher up than the previous one. The top row should end level with the base of the square roof ridge.

8 For the perch, I used a screwdriver that came with a self-assembly shelf kit; I simply drilled a hole the right size about 1 in. (2.5 cm) below the entrance hole and hammered the screwdriver in place. (Alternatively, you could use a 1¾-in. (4.5-cm) length of round dowel.) Drill a pilot hole in the square end of the front roof ridge, place a machine nut over the hole, and drive in a 1½-in. (40-mm) screw.

9 Remove the sharp edges from the roof flashing tile by curling the corners inward with a pair of needle-nose pliers. Bend the flashing tile in half over the roof ridge. Mark where each screw needs to go with a black marker pen. Using a 1/16-in. (1.5-mm) bit, drill pilot holes through the metal 1½ in. (4 cm) apart, starting ½ in. (12 mm) from the front end of the roof ridge and ¾ in. (2 cm) down from the top of the metal. Change to a ⅛-in. (3-mm) bit and drill pilot holes at the same points, then screw in ¾-in. (20-mm) wood screws.

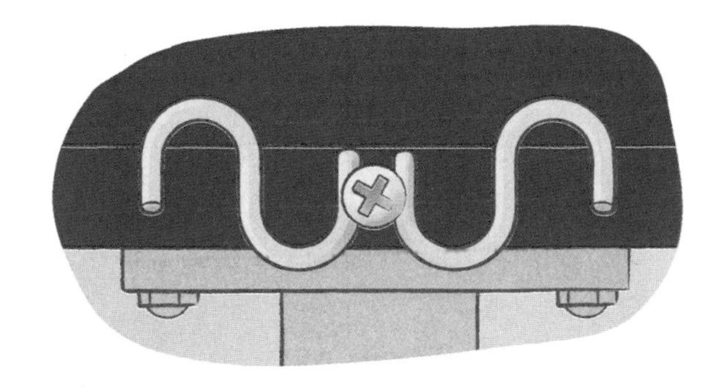

10 Starting from the center and working outward, using a ⅛-in. (3-mm) bit or one that is slightly smaller than the screw neck, drill a pilot hole for the first screw. Drill the screw in slightly, then place the end of an S-hook under each side and tighten the screw. Repeat all the way across the front panel.

11 Plug the entrance hole with wadded-up paper or tape, then varnish the painted areas of the birdhouse (see page 127). The varnish can go over the screws and S-hooks on the front, but you do not need to varnish the shims.

12 Mount the birdhouse on a hollow galvanized fence pipe, then plant the post in your garden, over 4-foot rebars (see page 135).

This beautifully ornate birdhouse is decorated with glass tiles and a metal roof and would suit a beach or waterfront home perfectly. It is easily attached to a tree, a fence, or the side of a wooden house by means of a back mount panel.

mosaic tiled birdhouse

Dark-Eyed Junco

materials

One 6 ft x 5½ in. 2½ in. (180 x 14 x 1.2 cm) dog-ear fence board
One 6 ft x 4 in. x 1 in. (180 x 10 x 2.5 cm) piece of lumber (timber)
Waterproof premium glue
1-in. (25-mm) finish nails or galvanized wire nails
Twelve ¾-in. (20-mm) wood screws
80-grit sandpaper
White oil-based primer paint
Light blue oil-based exterior spray paint
Light blue exterior craft paint
Two 12 x 12-in. (30 x 30-cm) sheets of small mosaic (meshed) glass tiles
Waterproof silicone glue
Gray grout
Six 1¼-in. (30-mm) exterior screws
Two 2-in. (50-mm) exterior screws
12 x 8 in. (30 x 20 cm) steel roof flashing (from home improvement store)
1¾-in. (4.5-cm) length of round dowel, ¼ in. (6 mm) in diameter
Dark blue glass knob, 1½ in. (4 cm) in diameter
Hummingbird pendant, approx. 2½ x 1¾ in. (6 x 4.5 cm) (optional)
2 in. (5 cm) heavy wire
Basic tool kit (see page 126)

finished size

Approx. 16 x 8 x 9 in. (40 x 20 x 23 cm)

interior dimensions

Floor area: 4¼ x 5½ in. (10.75 x 14 cm)
Cavity depth: 10 in. (25 cm)
Entrance hole to floor: 8 in. (20 cm)
Entrance hole: 1½ in. (40 mm) in diameter

cutting list

Front and back: 12 x 5½ in. (30 x 14 cm)—cut 2
Sides: 9¼ x 5½ in. (23.5 x 14 cm)—cut 2 (one side is reserved for the door)
Floor: 4¼ x 5½ in. (10.75 x 14 cm)—cut 1
Roof: 8 x 5½ in. (20 x 14 cm)—cut 2

1 Referring to the Basic Birdhouse on page 8, cut the birdhouse pieces from dog-ear fence board and shape them. Cut the peaks of the front and back panels at a 45° angle. Bevel the top edges of the side panels at a 45° angle. Do not bevel the roof panels. Cut ¼ in. (6 mm) diagonally off each corner of the floor panel. Cut a 1½-in. (40-mm) entrance hole in the front panel, 8½ in. (21.5 cm) from the bottom and centered on the width.

2 Assemble the body of the birdhouse, following steps 2–4 of the Basic Birdhouse on page 8. Glue and nail the two roof panels together, then glue and nail the roof to the birdhouse, flush with the back edge. The roof will overhang at the front to protect the entrance hole from rain.

3 Prepare the birdhouse for painting (see step 10 on page 9), then paint with white oil-based primer.

4 From lumber (timber), cut one back mount panel measuring 16 in. x 4 in. (40 x 10 cm). Cut off each end at a 45° angle, 1¼ in. (3 cm) from the edge, to make a dog-ear shape. Using a ⅛-in. (3-mm) bit, drill pilot holes 1 in. (2.5 cm) in from each side and ¾ in. (2 cm) down from the top. Drill in 2-in. (50-mm) exterior screws to attach the mount panel to a post or tree when the birdhouse is complete.

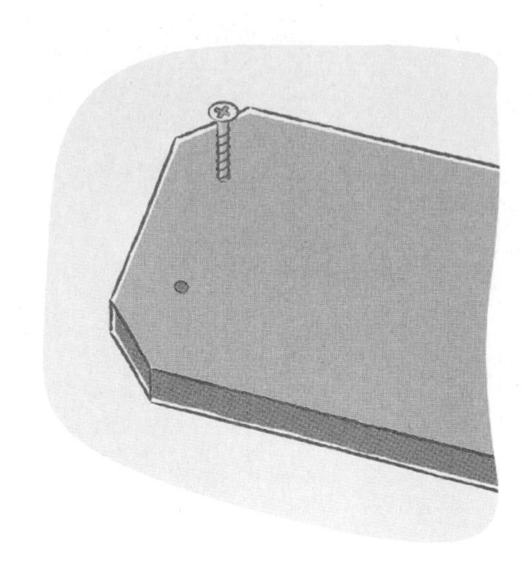

5 Paint the front and back of the mount panel with light blue oil-based exterior spray paint. Using a ⅛-in. (3-mm) bit, drill four pilot holes in the mount panel (see page 131), then drill in 1¼-in. (30-mm) screws to attach the back mount panel to the center of the back of the birdhouse.

6 Dry fit a sheet of mosaic (meshed) tile over the front, then cut to size with a utility knife by cutting through the mesh on the back of the tiles. Remove any tiles that hit the underside of the roof and save all the cut-off pieces for use later. Cut out a square of tiles around the entrance hole, and remove any tiles that cover up the screws attaching the door to the front panel.

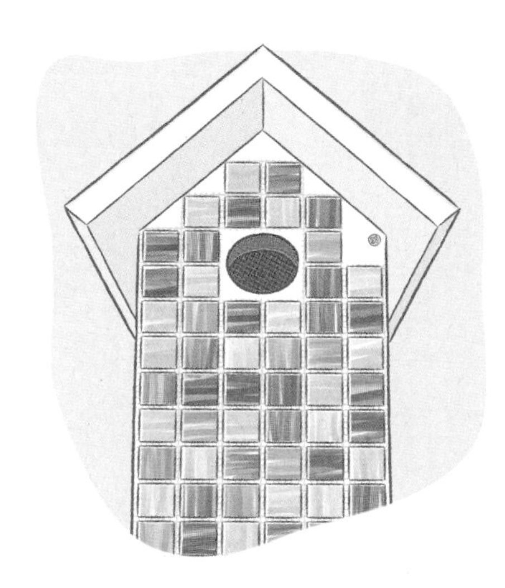

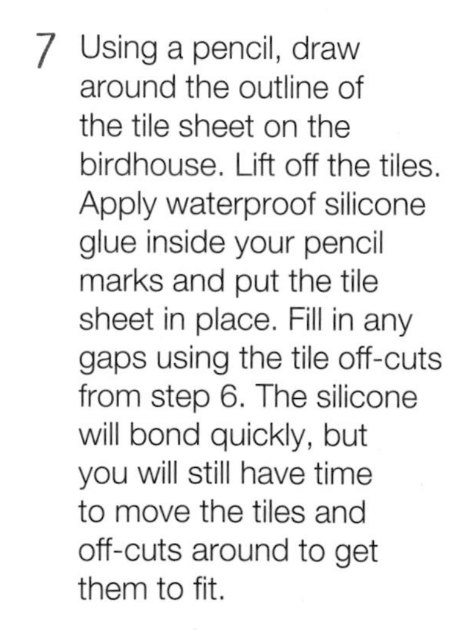

7 Using a pencil, draw around the outline of the tile sheet on the birdhouse. Lift off the tiles. Apply waterproof silicone glue inside your pencil marks and put the tile sheet in place. Fill in any gaps using the tile off-cuts from step 6. The silicone will bond quickly, but you will still have time to move the tiles and off-cuts around to get them to fit.

8 Repeat on the sides of the birdhouse, making sure that the door will still open and close properly when the tiles are in position. Remove tiles from where the door knob will go. Drill a $\frac{3}{16}$-in. (5-mm) hole and attach the door knob. Fill in any gaps using off-cuts from step 6.

9 Apply tiles to the front roof fascia; don't worry if you have space at the end, as you will fill it with grout later.

10 Put on gloves to protect your hands. Find the center of the steel flashing, place it on the edge of a table and press. Place the crease on the center of the roof and press it down flush with the roof panels.

11 Using a 1⁄16-in. (1.5-mm) bit, drill tiny indentations in the flashing along the lower edge of the roof, ¼ in. (6 mm) up from the edge and 4 in. (10 cm) apart. Change to a ⅛-in. (3-mm) bit and drill in ¾-in. (20-mm) wood screws to attach the flashing to the roof. Repeat along the ridge of the roof, spacing the screws 1¼ in. (3 cm) from the ridge, then repeat on the other side of the roof. Using flat-nose pliers, crimp the ridge of the roof, so that the flashing sits tight against the roof panels.

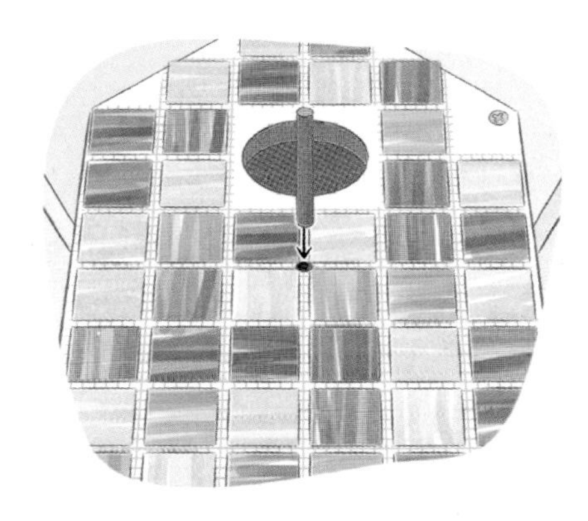

12 Using a ¼-in. (6-mm) bit, drill a hole in between two tiles about 1 in. (2.5 cm) below the entrance hole. Tap in a 1¾-in. (4.5-cm) length of round dowel for the perch.

13 Using a metal or rubber spatula, apply pre-mixed grout over the tiles, pressing it into each gap and around the entrance hole. Repeat on the sides and roof fascia, making sure you keep the door edges and opening free of grout. Leave for a few minutes (follow the instructions on the tub), then wipe with a damp sponge to smooth out the grout. Once the grout has dried firmly (about 15 minutes—but check the manufacturer's instructions), wipe off any film that has formed on the surface with a damp sponge.

14 If you're adding a hummingbird pendant, decide where you want it to go and drill a ⅛-in. (3-mm) hole between two tiles at this point. Thread a length of wire through the loop on the back of the pendant and twist the ends together. Using a toothpick, apply a tiny dab of silicone glue to the inside of the hole, then press the wires onto the glue to secure, making sure they do not stick through to the inside of the birdhouse.

15 Paint the base of the birdhouse in light blue exterior craft paint, to match the back mount panel. Apply the same color in any other places where the primer shows through. You do not need to varnish this birdhouse, as the grout will have waterproof sealant in it.

With its white squares on a dark purple base, this birdhouse has a definite retro feel! Using a checkerboard stencil is a fun and easy way to dress up any project.

checkerboard birdhouse

materials

One 6 ft x 5½ in. x ½ in. (180 x 14 x 1.2 cm) dog-ear fence panel
One 6 ft x 7½ in. x ½ in. (180 x 19 x 1.2 cm) dog-ear fence panel
Waterproof premium glue
1-in. (25-mm) finish nails or galvanized wire nails
1¾-in. (4.5-cm) length of 5⁄16-in. (8-mm) round dowel
Four ¾-in. (20-mm) machine screws
Six 1¼-in. (30-mm) exterior screws
Two 2½-in. (60-mm) exterior screws
Paintable wood-filler putty
80-grit sandpaper
Stencil with checkerboard design
Masking tape
Dark purple oil-based exterior spray paint
White or yellow exterior craft paint
Water-based exterior varnish
Basic tool kit (see page 126)

finished size

Approx. 6½ x 6¼ x 7½ in. (16.5 x 16 x 19 cm)

interior dimensions

Floor area: 4¼ x 5½ in. (11 x 14 cm)
Cavity depth: 8 in. (20 cm)
Entrance hole to floor: 6 in. (15 cm)
Entrance hole: 1½ in. (40 mm) in diameter

cutting list

Back panel: 12 x 5½ in. (30 x 14 cm)—cut 1
Front panel: 9 x 5½ in. (23 x 14 cm)—cut 1
Side panels: 11 x 5½ in. (29 x 14 cm)—cut 2
Floor: 4¼ x 5½ in. (11 x 14 cm)
Platform roof: 9 x 7½ in. (23 x 19 cm)—cut 1
Back plate: 11 x 7½ in. (28 x 19 cm)—cut 1

1 Referring to the Basic Birdhouse on page 8, cut the birdhouse pieces from dog-ear fence board. Bevel the top edges of the front and back panels at 22.5°. Cut the top edges of the side panels at a 22.5° angle. Cut a 1½-in. (40-mm) entrance hole, 6 in. (15 cm) from the bottom of the front panel and centered on the width. Drill a hole and insert a length of round dowel for the perch 1 in. (2.5 cm) below the entrance hole (see step 9 on page 9).

2 Assemble the body of the birdhouse, following steps 2–4 of the Basic Birdhouse on page 8 as a guide, but note that the front panel is the door on this birdhouse. Glue and nail the roof panel in place, overhanging the back panel by ½ in. (1 cm).

3 Glue and screw in the back plate, with the top edge flush with the back panel and the sides overhanging by 1 in. (2.5 cm), for extra strength. Remove the door. Using a ⅛-in. (3-mm) bit, drill pilot holes (see page 131) through the back panel and back plate for mounting. Drill in 2½-in. (60-mm) screws. Replace the door.

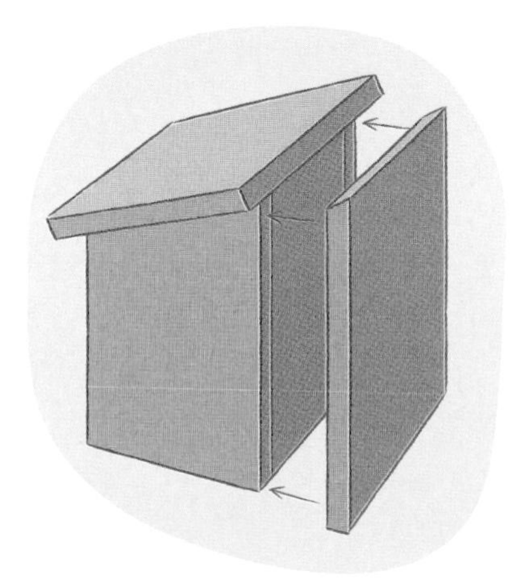

4 Using a ⁵⁄₁₆-in. (8-mm) bit, drill ventilation holes in the side panels just under the eaves, sloping upward to prevent rain from getting in.

5 Prepare the birdhouse for painting (see step 10 on page 9). Paint the whole birdhouse dark purple, then leave to dry. Tape a checkerboard stencil to the front of the birdhouse. Using a stiff-bristled brush or a stencil brush, dry brush white or yellow exterior craft paint through the stencil to create the pattern. Repeat on the side and front edges of the roof.

The perfect fit for tiny cavity-nesting birds, with two full bedrooms on both ends and double doors on the side, this birdhouse is a farmhouse barn painted in Apple Red. Hand-painted flowers on the side give it a fun, country feel.

barn birdhouse

materials

Three 6 ft x 5½ in. x ⅝ in. (180 x 14 x 1.5 cm) cedar dog-ear fence boards
One 6 ft x 7½ x ⅝ in. (180 x 19 x 1.5 cm) cedar dog-ear fence board
Square ¾ x ¾ in. (20 x 20 mm) dowel
Waterproof premium glue
Wood putty
80-grit sandpaper
1-in. (25-mm) and 1¼-in. (30-mm) finish nails or galvanized wire nails
Ten 1¼-in. (30-mm) exterior screws
Weatherproof 30-minute clear silicone
Red and gray exterior spray paint
"Found Objects" charms (I used gear findings)—available from craft stores
Cotton swabs (buds)
Two white ceramic drawer knobs
Two nuts and washers
Basic tool kit (see page 126)
White, yellow, green, and dark green craft paint

finished size

Approx. 17½ x 13 x 14 in. (44.5 x 33 x 35.5 cm)

interior dimensions

Interior dimensions (for each "room" of barn birdhouse)
Floor area: 4¼ x 5½ in. (10.8 x 14 cm)
Cavity depth: 12 in. (30 cm)
Entrance hole to floor: 9 in. (23 cm)
Entrance hole: 1⅛ in. (28.5 mm) in diameter

cutting list

Main birdhouse
Front and back: 12 x 5½ in. (30 x 14 cm)—cut 4
Sides: 9 x 5½ in. (23 x 14 cm)—cut 4
Floor: 4¼ x 5½ in. (10.8 x 14 cm)—cut 2
Roof: 16½ x 5½ in. (42 x 14 cm)—cut 2

Side birdhouse
Sides: 7¾ x 5½ in. (19.5 x 14 cm)—cut 2, with 1 short end cut at a 30° angle
Back: 6¾ x 5½ in. (17 x 14 cm)—cut 1
Floor: 4½ x 5½ in. (11.5 x 14 cm)—cut 1
Front: 4½ x 5½ in. (11.5 x 14 cm)—cut 1
Roof: 7½ x 7½ in. (19 x 19 cm)—cut 1, with 1 short end beveled at 30°

1 Cut all four front and back pieces for the main birdhouse at a 45° angle on both sides to create a peak (see page 8). On two of these pieces, cut a 1⅛ in. (28.5-mm) entrance hole 3¼ in. (8 cm) down from the peak (these will be the front panels).

2 Apply glue to the left side panel and place a front panel on top. Nail together. Glue and nail the back to the side panel.

3 Dry fit the floor to make sure the door will fit flush to the edges. Then glue and nail the floor in place.

4 Repeat steps 2 and 3 with the remaining panels—but this time attach the side panel to the *right-hand* edge of the front. This ensures that the doors will be on the same side when the two birdhouses are sandwiched together back to back.

5 Attach the remaining side panels (the doors) before joining the houses together. Use a ⁵⁄₁₆-in. (8-mm) bit to countersink the screws (see page 131). The screws on the back panels must be flush with the wood in order for the two birdhouses to be joined together in the next step. Make sure both doors open properly.

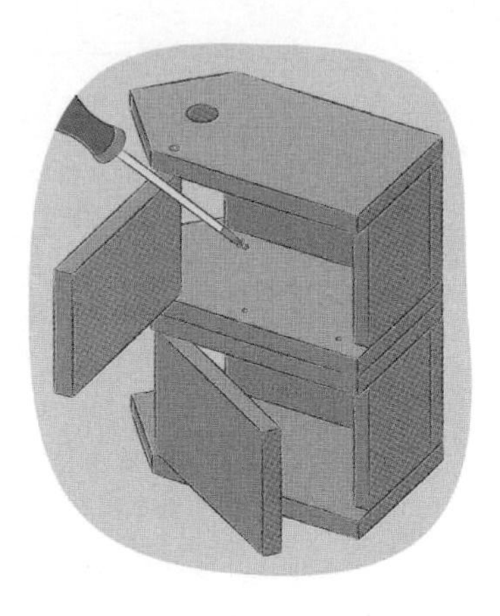

6 Apply glue to both back panels, then carefully line the two birdhouses up together. Drill pilot holes (see page 131) at an angle on the top and bottom back edges, then drive in 1¼-in. (30-mm) screws. Nail the edges by the doors and around the peak. Turn the birdhouse over and drill pilot holes and insert screws from the other side, in order to join the birdhouses really securely.

7 Apply glue to the roof peaks. Line up the first roof panel level with the peaks, with the same amount (1⅝ in./4.2 cm) overhanging each front, and nail in place. Repeat on the other side, but do not overlap the roof panels; this will leave a V-shaped gap between the roof panels that will be filled with a square dowel.

Before you attach the roof panels, mark where the peaks of the roof will sit beneath them so that you can be sure you're nailing through both layers.

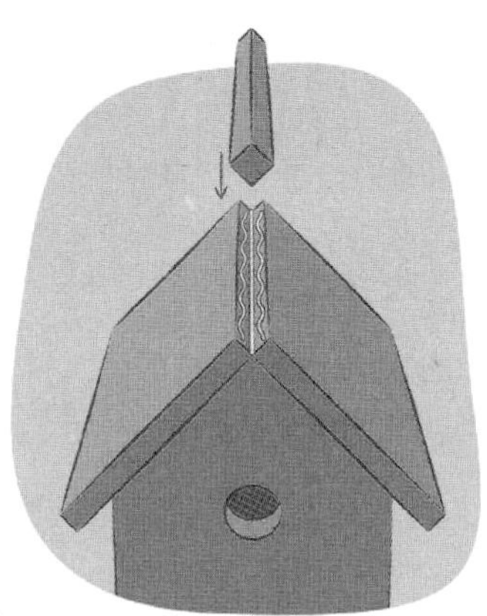

8 Cut the square dowel to 17½ in. (44.5 cm) long. Apply a bead of silicone to seal the tiny gap at the bottom of the "V" in the roof. Before the silicone dries, apply glue down each side of the "V." Insert the dowel, leaving a ½-in. (12-mm) overhang at each end, and nail in place with 1¼-in. (30-mm) nails.

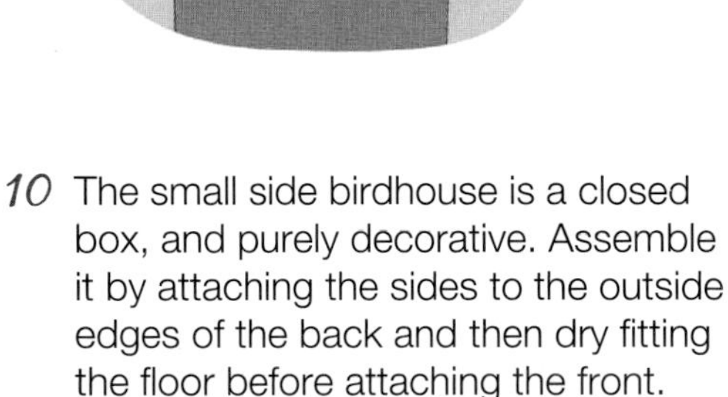

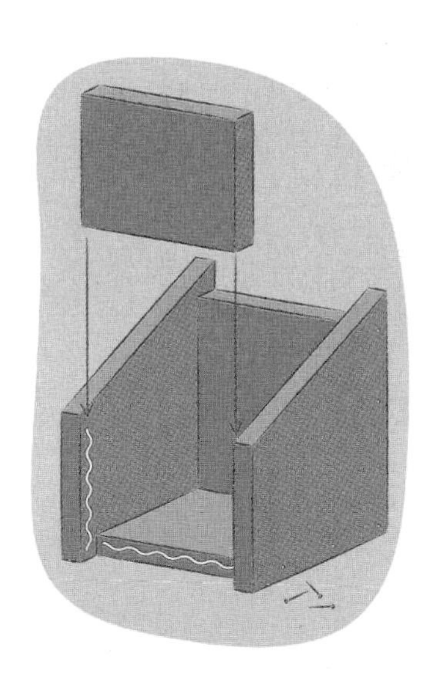

9 Paint the body of the birdhouse red and the roof dark gray and let dry.

10 The small side birdhouse is a closed box, and purely decorative. Assemble it by attaching the sides to the outside edges of the back and then dry fitting the floor before attaching the front. Paint red and let dry.

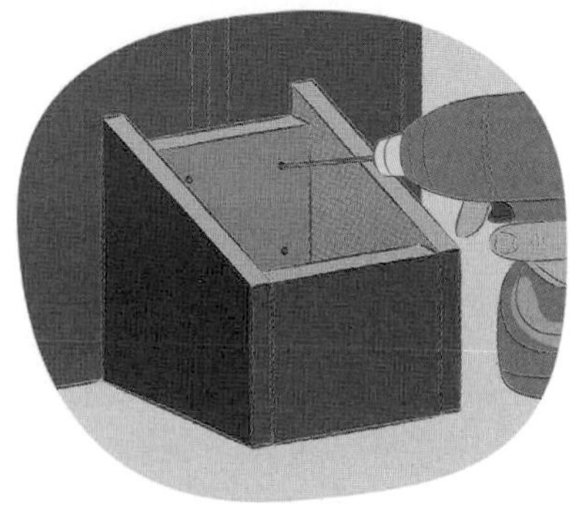

11 Apply glue to the back of the small house. Center it on the back of the barn birdhouse (the side without the doors!). Drill pilot holes in the back of the small house and insert 1¼-in. (30-mm) screws to join the houses together.

12 Paint the roof of the small house gray and let dry. Apply glue to the top edge of the small house, then slide the beveled edge of the roof up against the barn birdhouse, with the same amount overhanging on each side. Nail in place.

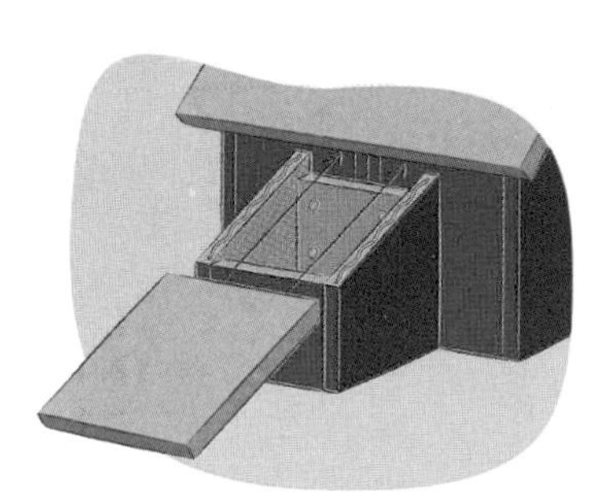

13 Cut six 5½-in. (14-cm) and two 2¼-in. (5.7-cm) lengths of square ¾-in. (20-mm) dowel. Paint gray and let dry. Apply glue and, using 1¼-in. (30-mm) nails, attach one long strip across the bottom of the barn birdhouse front, one 3¾ in. (9.5 cm) above it, and one to the small house, 2¼ in. (5.7 cm) from the bottom. Attach a shorter strip vertically to the small house, 1½ in. (4 cm) from the side. Repeat on the opposite side.

14 Cut two more pieces of dowel approx. 5½ in. (14 cm) long, with one end of each at a 45° angle. Place them diagonally across the two dowels on the barn birdhouse and mark where to cut at a 45° angle to complete the "Z." Paint gray and let dry, then glue and nail in place.

Once you've attached all the square dowels, change back to using 1-in. (25-mm) nails.

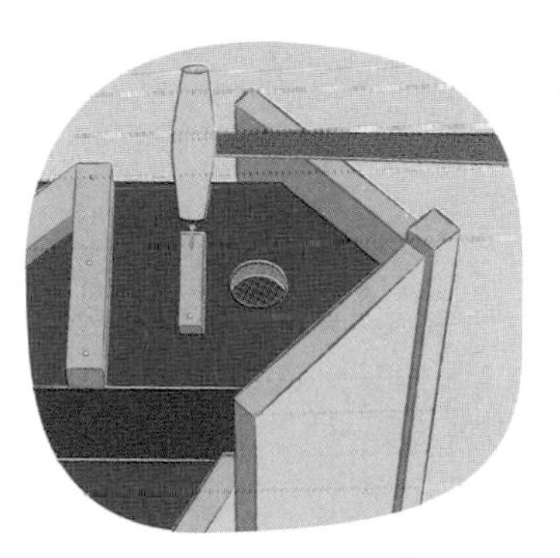

15 From scrap fence board, cut a strip 5½ x ½ in. (14 cm x 12 mm). Cut it in half widthwise to give two pieces 2¾ in. (7 cm) long. Paint gray and let dry. Glue and nail them centrally ¾ in. (2 cm) under the entrance holes.

16 Using a stiff paintbrush, drybrush white paint over all the gray dowels and roof, leaving some of the gray showing through.

17 Attach gear findings above the "T-bars" on both sides of the small house by applying a dab of silicone with a cotton swab (bud).

18 Find the center of each door. Using a 3⁄16-in. (5-mm) bit, drill a hole 2 in. (5 cm) above the bottom edge of each clean-out door, and insert a door knob, securing them on the inside with a nut and washer.

19 Hand paint flowers onto the unadorned side of the small house (see page 139).

As the saying goes, "Everyone's trash is someone else's treasure"—and that's what this planter-box birdhouse is all about. I had found an old rusted faucet valve in my garden that I had put away for a rainy day, and gathered other small trinkets from craft stores and our construction business. With this project, anything goes—just use your imagination!

planter-box birdhouse

materials

Two 6 ft x 5½ in. x ½ in. (180 x 14 x 1.2 cm) dog-ear fence panels
Waterproof premium glue
1-in. (25-mm) finish nails or galvanized wire nails
25–30 x 1¼-in. (30-mm) exterior screws
1¾-in. (4.5-cm) length of 5/16-in. (8-mm) round dowel
Paintable wood-filler putty
80-grit sandpaper
Black oil-based exterior spray paint
Mid-green oil-based exterior spray paint
Decorative items of your choice—for example, pine bark for the roof, an old fork for the door pull, driftwood sticks, moss, hooks and door plates from your local hardware or craft store, an old rusty faucet (tap), decorative buttons
Multi-purpose waterproof adhesive and small screws to attach the decorations
Cream exterior craft paint
Water-based exterior varnish
Basic tool kit (see page 126)

finished size

Birdhouse: 21½ x 6⅜ x 5½ in. (54.5 x 16.25 x 14 cm)
Planter box: 9 x 22½ in. (23 x 57 cm)

interior dimensions

Floor area: 4¼ x 5½ in. (11 x 14 cm)
Cavity depth: 10 in. (25 cm)
Entrance hole to floor: 8 in. (20 cm)
Entrance hole: 1¼ in. (32 mm) in diameter

cutting list

Planter box

Adapt the measurements to fit your chosen plant containers if necessary.
Base: 21 x 5½ in. (53 x 14 cm)—cut 1
Side panels: 21 x 5½ in. (53 x 14 cm)—cut 2
End panels: 5½ x 5½ in. (14 x 14 cm)—cut 2

Birdhouse

Front: 18 x 5½ in. (45.75 x 14 cm)—cut 1
Back: 13¼ x 5½ in. (33.5 x 14 cm)—cut 1
Sides: 10½ x 5½ in. (26.5 x 14 cm)—cut 2 (one side is reserved for the door)
Bottom roof: 5 x 5½ in. (20 x 14 cm)—cut 2
Top roof: 5¾ x 5½ in. (14.5 x 14 cm)—cut 2
Floor: 4¼ x 5½ in. (11 x 14 cm)—cut 1

1 Glue and nail the side panels of the planter box to the outside of the base panel, then drill pilot holes (see page 131) and insert 1¼-in. (30-mm) screws for extra security. Attach the end panels to the outside of the side panels in the same way. Drill 5/16 in. (8-mm) drainage holes in the bottom of the planter box.

2 Paint the inside of the box in black oil-based exterior paint and leave to dry.

3 Referring to the Basic Birdhouse on page 8, cut and shape the birdhouse pieces from dog-ear fence board. To determine where the entrance hole will go, place the front panel on the front of the planter box. Using a speed square, mark a line on the back of the front panel, level with the top edge of the planter box. Cut a 1¼-in. (32-mm) entrance hole, 8½ in. (21.5 cm) above the pencil line and centered on the width.

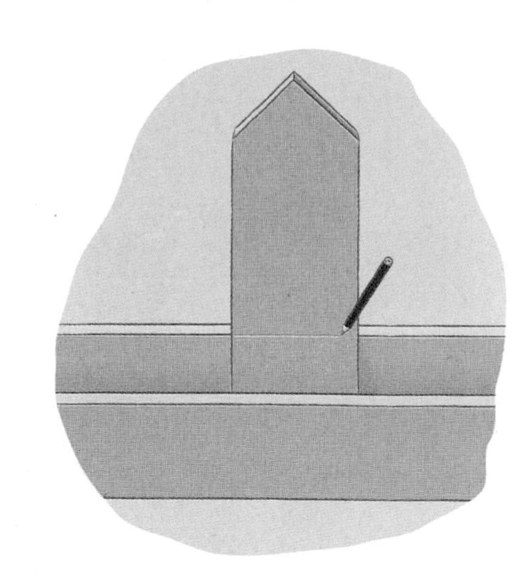

4 Assemble the birdhouse and drill a hole for the dowel perch, following steps 1–9 of the Basic Birdhouse on pages 8–9, aligning the front, side and back panels at the top; the front panel will overhang the others at the base, so that you can attach it to the planter box.

5 Prepare the planter box and birdhouse for painting (see step 10 on page 9). Turn the planter box upside down to avoid getting paint on the inside, then paint the planter box and birdhouse mid-green.

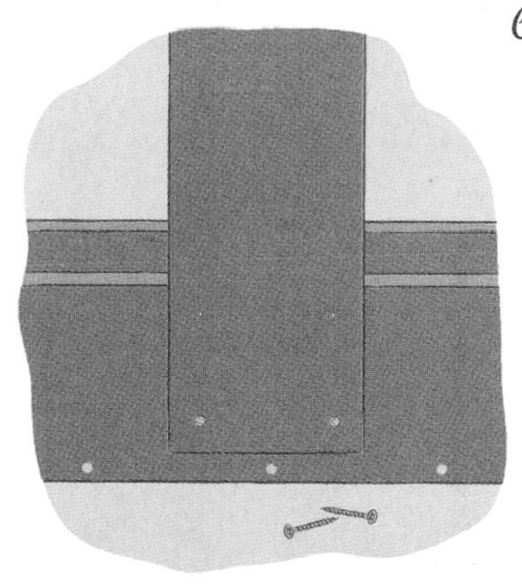

6 Using a ⅛-in. (3-mm) bit, drill four pilot holes (see page 131) at the base of the front panel, then insert 1¼-in. (30-mm) screws to attach the birdhouse to the front of the planter box.

7 Attach your chosen embellishments wherever you choose. You can give metal items, such as door plates and hooks, the patina of old bronze by lightly dabbing on two shades of green craft paint.

8 Dry brush the outside of the birdhouse and planter box with cream exterior craft paint, allowing some of the underlying green to show through (wipe off excess paint with a rag if necessary). Plug the entrance hole with wadded-up paper or tape, then varnish the outside of the birdhouse (see page 127).

9 Place the planter on a shelf or attach it to a 4 x 4-in. (10 x 10-cm) post, screwing decorative corbels to the plant base and the sides of the post.

Winter Wren

Who doesn't like the 1960s era?! It's fun, fabulous, and colorful. This unique birdhouse is functional, yet retro with clean lines. The bottom drawer pulls out completely to remove nesting materials once the season is over. It's sure to be an eye-catcher.

dresser drawer birdhouse

materials

One 6ft x 5½ in. x ⅝ in. (180 x 14 x 1.5 cm) cedar dog-ear fence board
Two 36-in. (1-m) square ¾ x ¾-in. (20 x 20-mm) dowels
One 48-in. (1.2-m) round ¾-in. (20-mm) dowel
Waterproof premium glue
Wood putty
80-grit sandpaper
1-in. (25-mm) and 1¼-in. (30-mm) finish nails or galvanized wire nails
Lime green exterior spray paint
Blue, orange, and yellow craft paint
Basic tool kit (see page 126)

finished size

Approx. 13½ x 8 x 8½ in. (34 x 20 x 21.5 cm)

interior dimensions

Floor area: 4¼ x 5½ in. (10.8 x 14 cm)
Cavity depth: 9 in. (23 cm)
Entrance hole to floor: 7 in. (18 cm)
Entrance hole: 1¼ in. (30 mm) in diameter

cutting list

Front, back, and sides: 10 x 5½ in. (25 x 14 cm)—cut 4
Base: 5½ x 4¼ in. (14 x 10.8 cm)—cut 1
Drawer inner base: 5½ x 4³⁄₁₆ in. (14 x 10.6 cm)—cut 1
Drawer sides: 5½ x ⅝ in. (14 x 1.5 cm)—cut 2
Roof: 8 x 5½ in. (20 x 14 cm)—cut 2, then cut 1 in half lengthwise

1 Cut a 1¼-in. (30-mm) entrance hole in the front panel, 2 in. (5 cm) down from the top edge (see page 130).

2 Mark the front into three sections, the top and bottom sections 3¾ in. (9.5 cm) high and the center 2½ in. (6 cm) high. Cut into three sections for the drawer fronts.

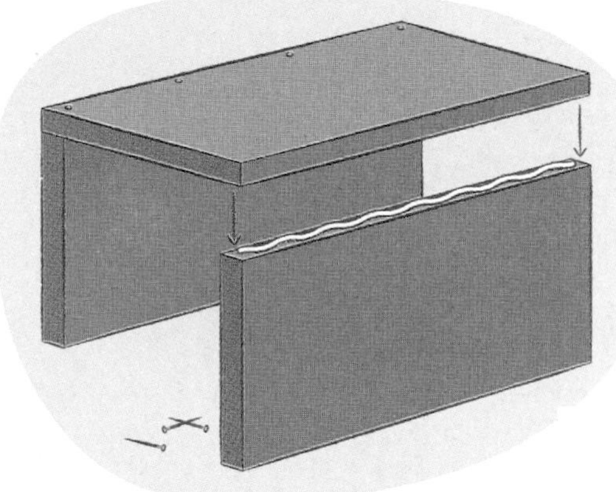

3 Glue and nail the back to the edges of the side pieces.

4 Glue and nail the top front section (the one with the entrance hole) in place, flush with the top and side edges. Glue and nail the center front section in place, ⅛ in. (3 mm) below it (this section does not open).

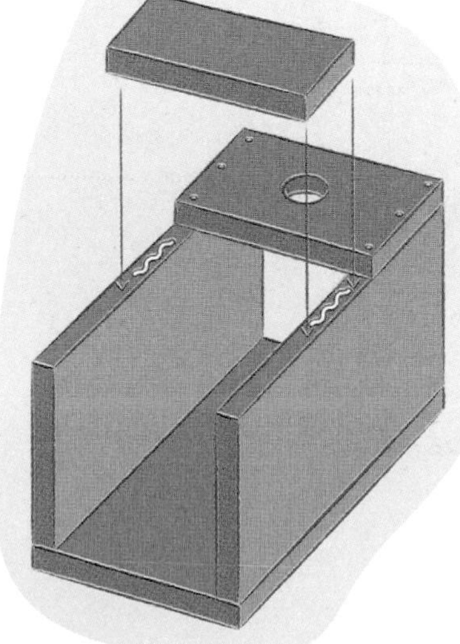

5 Glue and nail the base in place, flush with the edges of the back and sides.

6 Slide the drawer base inside the carcass, apply glue to the front edge, and position the lower section of the front (the drawer front) on top, flush with the bottom edge and sides of the carcass. Nail in place; there will be a ⅛-in. (3-mm) gap at the top.

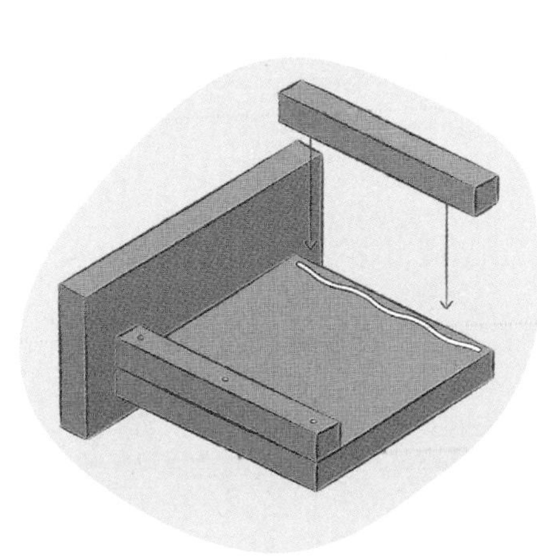

7 Remove the drawer, then glue and nail the drawer sides to the base, flush with the edges.

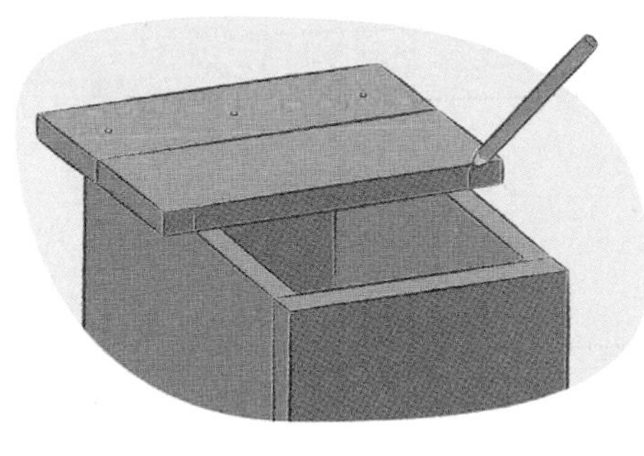

8 Apply glue to the top front edge of the birdhouse and approx. 2¾ in. (7 cm) along the side edges. Place a halved roof panel on top, overhanging by 1⅜ in. (3 cm) at the sides and by 1¼ in. (2.8 cm) at the front, and nail in place. Place the halved back roof panel flush with the front, then mark where the sides meet the roof.

9 Apply glue to the back edge of the birdhouse, line up the back panel markings with the back corners with a ¼-in. (6-mm) overhang, and nail in place.

10 Apply glue to the inner edges of the bottom roof panels, center the top roof on top, lining up the side edges with those of the bottom roof panels, and nail in place.

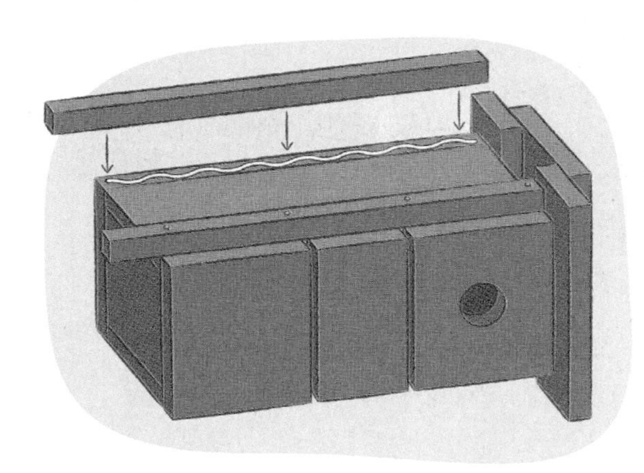

After cutting all four legs, place them all together with the dowels back against a speed square on the chop saw, then cut all the ends at once to ensure they are all the same length.

11 Cut four square dowels 11½ in. (29 cm) long. Using 1¼-in. (30-mm) nails, attach one to each side of the front, ⅝ in. (1.5 cm) back from the front edge. Attach the back legs flush with the back edge.

12 Cut two 1½-in. (4-cm) lengths of round dowel for the drawer knobs. Glue and nail them to the center of the fake middle drawer and the bottom pull-out drawer, using 1¼-in. (30-mm) nails. Cut a 4-in. (10-cm) length of dowel and glue and nail it to the center of the top roof.

13 Paint the birdhouse in your chosen color. I spray painted the whole piece lime green, then painted the bottom roof and center drawer face blue, the knobs and top round dowel yellow, and the legs orange. Finally I added small orange circles around the entrance hole with the end of a paintbrush.

Upcycling is the process of converting waste materials or useless products into new materials or products of better quality or a higher environmental value. In this birdhouse, I used wine corks from all the vintages of wine that I love. The birdhouse also has a cognac cork for the chimney and a cork for the doorknob.

upcycled wine cork birdhouse

materials

One 6 ft x 5½ in. x ½ in. (180 x 14 x 1.2 cm) dog-ear fence board
Waterproof premium glue
1-in. (25-mm) finish nails or galvanized wire nails
½-in. (12-mm) finish nails or galvanized wire nails
Five 1⅝-in. (40-mm) exterior screws
Two 1¼-in. (30-mm) exterior screws
Cognac cork
1-in. (25-mm) EMT two-hole pipe clamp (clip)
Two ¾-in. (20-mm) wood screws
Paintable wood-filler putty
80-grit sandpaper
Green oil-based exterior spray paint
Brown exterior craft paint
Water-based exterior varnish
Approx. 95 wine corks (some will be cut in half)
Small bunch of artificial grapes
30 in. (75 cm) heavy wire tie
8 sticks, 2–7 in. (5–18 cm) in length
Green moss
Basic tool kit (see page 126)

finished size

Approx. 12 x 11 x 7 in. (30 x 28 x 18 cm)

interior dimensions

Floor area: 4¼ x 5½ in. (11 x 14 cm)
Cavity depth: 9 in. (23 cm)
Entrance hole to floor: 6 in. (15 cm)
Entrance hole: 1½ in. (40 mm) in diameter

cutting list

Front and back: 9¼ x 5½ in. (23.5 x 14 cm)—cut 2
Sides: 6¼ x 5½ in. (16 x 14 cm)—cut 2 (one side is reserved for the door)
Bottom roof: 5 x 5½ in. (13 x 14 cm)—cut 2
Top roof: 5¾ x 5½ in. (14.5 x 14 cm)—cut 1
Floor: 4¼ x 5½ in. (11 x 14 cm)—cut 1
Porch: 10½ x 5½ in. (26.5 x 14 cm)—cut 1 from dog-ear end of panel

1 Referring to the Basic Birdhouse on page 8, cut and shape the birdhouse pieces from dog-ear fence board. Cut a ½-in. (40-mm) entrance hole, 6½ in. (16.5 cm) from the bottom of the front panel and centered on the width.

2 Assemble the birdhouse, following steps 2–8 of the Basic Birdhouse on pages 8–9.

3 Drill a ⅛-in. (3-mm) pilot hole just below the roof ridge to attach the cork chimney. Drill in a 1⅝-in. (40-mm) exterior screw from the underside of the roof, protruding on the outside of the roof.

4 Using a hand saw, cut the cork for the chimney at a 45° angle. Apply waterproof glue under and around the cork, then twist it down onto the screw until it is flush with the roof. Wipe off any excess glue.

5 Attach a 1-in. (25-mm) EMT two-hole pipe clamp (clip) to the apex of the roof for a hanging loop (see page 132).

6 Prepare the birdhouse for painting (see step 10 on page 9). Spray the outside of the birdhouse and both sides of the bottom platform with green paint.

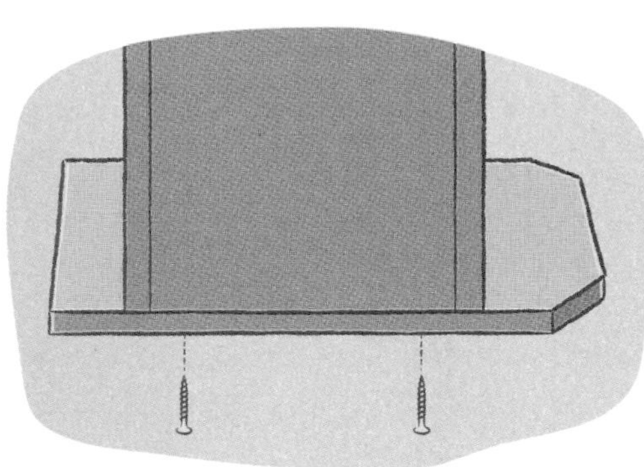

7 Drill ⅛-in. (3-mm) pilot holes through the bottom of the porch, then attach the birdhouse to the porch with 1¼-in. (30-mm) screws, making sure that the back of the porch extends 1½ in. (4 cm) beyond the back panel of the birdhouse. The dog-ear end of the porch should be at the front of the birdhouse.

8 Remove the door of the birdhouse. Drill a ⅛-in. (3-mm) pilot hole from the back of the door, then drill in a 1¼-in. (30-mm) screw. Twist a wine cork onto the screw for the door knob.

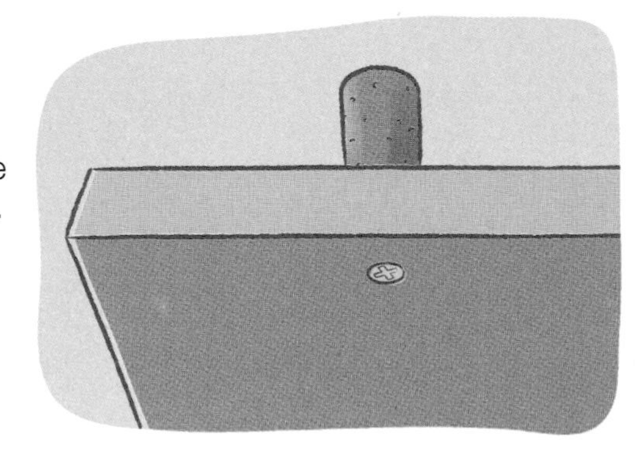

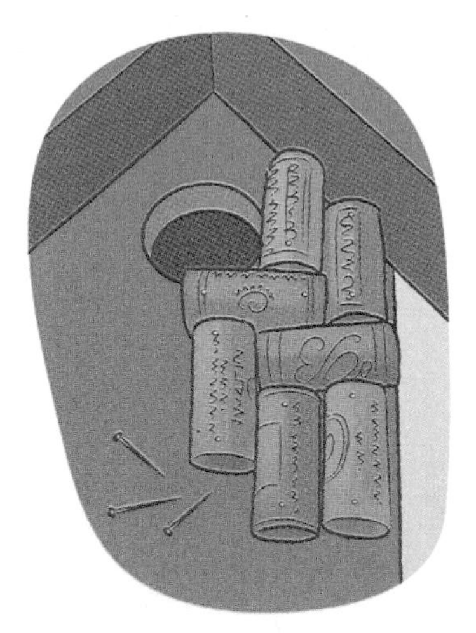

9 Using waterproof glue, start adding full corks to the front of the birdhouse, then nail in place ¼ in. (6 mm) from each end of each cork. Use a hand saw to cut corks at angles or trim them to fit in tight spots. If small bits of cork are protruding or causing an obstruction, trim with utility scissors. Gently hammer cork trimmings into any gaps.

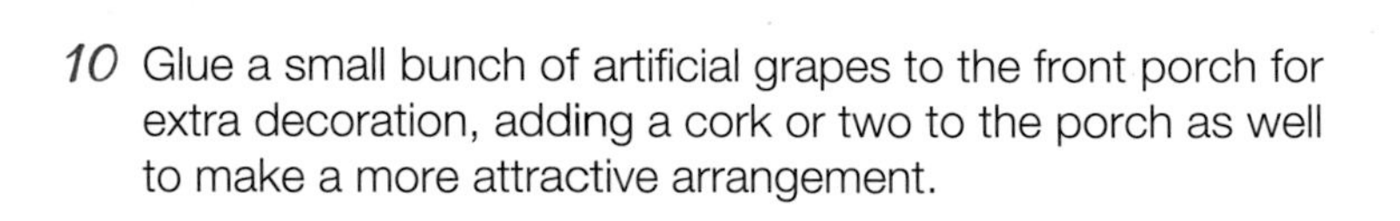

10 Glue a small bunch of artificial grapes to the front porch for extra decoration, adding a cork or two to the porch as well to make a more attractive arrangement.

11 Glue and nail half corks to the door of the birdhouse, making sure you don't nail corks to the side edges of the other panels in order to allow the door to open. Keep checking the door to make sure glue has not seeped between the crack of the door onto the edge or that you've accidentally nailed the door shut. Glue and nail half corks to the side opposite the door.

12 Glue and nail six full corks vertically to the porch on the back of the birdhouse with a seventh cork horizontally on top. Paint the roof with diluted brown exterior craft paint (1 part paint to 3 parts water), to allow the green color to bleed through. Let dry. Plug the entrance hole with wadded-up paper or tape, then varnish the birdhouse (see page 127). Let dry.

13 Using a ⅛-in. (3-mm) bit, drill three holes through a 6-in. (15-cm) piece of driftwood, in the center of the wood and 1½ in. (4 cm) from each end. Thread wire through the holes, as shown. Loop the hanger end of the wire through the top hole again to create a loop for hanging, then twist with pliers to keep the wire tight and secure.

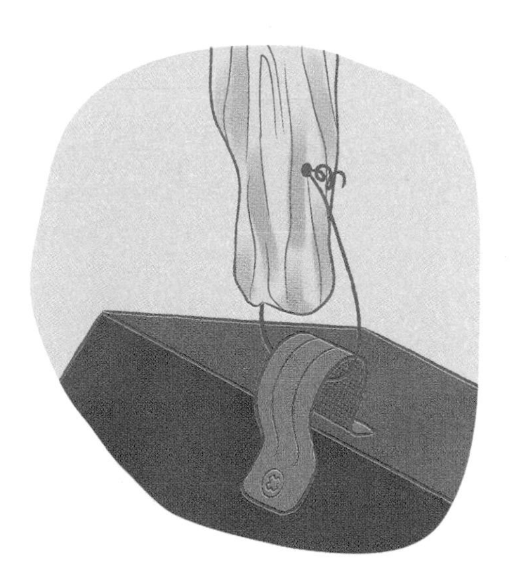

14 Loop the other end of the wire through the pipe clamp (clip), then form into a loop, as in the previous step.

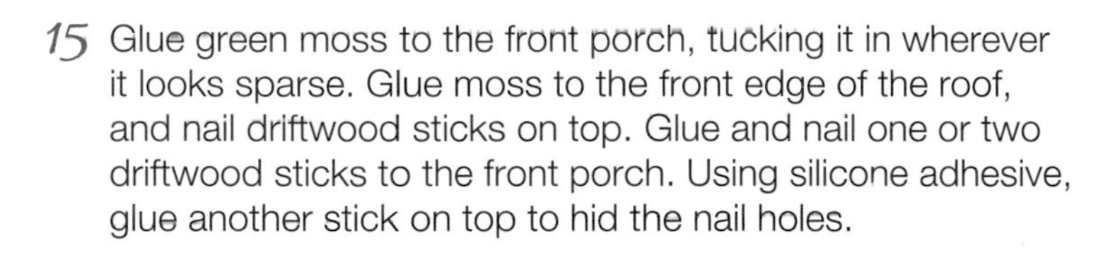

15 Glue green moss to the front porch, tucking it in wherever it looks sparse. Glue moss to the front edge of the roof, and nail driftwood sticks on top. Glue and nail one or two driftwood sticks to the front porch. Using silicone adhesive, glue another stick on top to hid the nail holes.

A simple home for small cavity-nesting birds, this birdhouse can easily be hung from an iron garden pole. Although the box seems small, the area in which the birds will build their nest is the perfect size. Ventilation holes on the back and both sides also act as drainage. The side door lifts for easy clean-out, with the screw keeping the door secure while birds are caring for their young.

hanging cube birdhouse

materials

- One 6 ft x 5½ in. x ½ in. (180 x 14 x 1.2 cm) dog-ear fence panel
- Waterproof premium glue
- 1-in. (25-mm) finish nails or galvanized wire nails
- Three 1¼-in. (30-mm) exterior screws
- 80-grit sandpaper
- Paintable wood-filler putty
- 1¾-in. (4.5-cm) length of round dowel, 5⁄16 in. (8 mm) in diameter
- Lime-green oil-based exterior spray paint
- Dark brown and dark green exterior craft paint
- Water-based exterior varnish
- Two large chain links
- 5-in. (12-cm) length of chain
- 1-in. (25-mm) EMT two-hole pipe clamp (clip)
- Two ¾-in. (20-mm) wood screws
- Basic tool kit (see page 126)

finished size

Approx. 10¾ x 8¾ x 5½ in. (27.5 x 22.25 x 14 cm)

interior dimensions

Floor area: 6 x 4 in. (15 x 10 cm)
Cavity depth: 6 in. (15 cm)
Entrance hole to floor: 4 in. (10 cm)
Entrance hole: 1¼ in. (32 mm) in diameter

cutting list

Front and back: 5½ x 5½ in. (14 x 14 cm)—cut 2
Roof panel 1: 5½ x 7½ in. (14 x 19 cm)—cut 1
Roof panel 2: 5½ x 7¾ in. (14 x 19.5 cm)—cut 1
Left side panel: 4¾ x 4 in. (12 x 10 cm)—cut 1
Right side panel (door): 5¼ x 4 in. (13.25 x 10 cm)—cut 1

Great Tit

1 Cut the birdhouse pieces from dog-ear fence panel. Turn the front panel so that you're looking at it as a diamond shape. Cut a 1¼-in. (32-mm) entrance hole, 1¾ in. (32 mm) from the top point and centered on the width.

2 Glue and nail the long roof panel over the top of the short roof panel. Glue and nail the back panel flush to the inside edge of the roof panels, nailing from the top of the roof panels. At each corner where the back panel meets the roof panel, measure 4 in. (10 cm) up the roof panel and make a pencil mark to show where the front panel will attach to the underside of the roof.

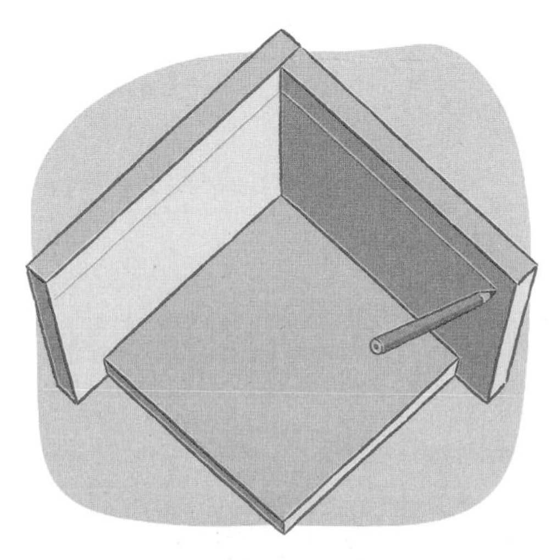

3 Glue and nail the front panel in place, nailing from the top of the roof panels. The roof will slightly overhang the front of the birdhouse.

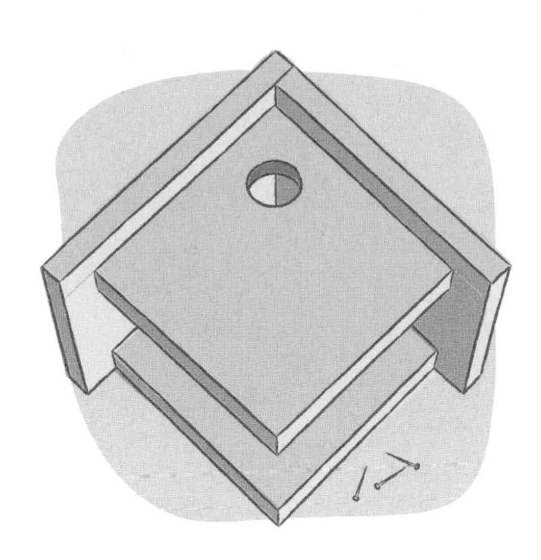

4 Glue and nail the left side panel to the inside edges of the front and back panels. Using a ⅛-in. (3-mm) bit, drill pilot holes (see page 131) on the sides of the front and back panels, then attach the right side panel (the door) using 1¼-in. (30-mm) exterior screws. As this door is on a downward slope, you will also need to drill another pilot hole from the base of the right-side panel and screw in a 1¼-in. (30-mm) exterior screw so that the door can't open during the nesting season; slightly unscrew this screw to remove the nest when necessary.

5 Drill a hole in the front panel about 1 in. (2.5 cm) below the entrance hole and gently tap in the round dowel for the perch (see step 9 on page 9). Using a 3⁄16-in. (5-mm) bit, drill holes just under the eaves on both sides and the back panel for ventilation and drainage, sloping upward to prevent water from dripping in.

6 Prepare the birdhouse for painting (see step 10 on page 9). Prime with gray oil-based exterior spray paint. Leave to dry, then paint with lime-green oil-based exterior spray paint. Leave to dry.

7 Apply dark brown exterior craft paint to the roof. While still slightly damp, apply dark green in uneven strips. Leave to dry. Plug the entrance hole with wadded-up paper or tape, then varnish the exterior of the birdhouse (see page 127).

8 Attach a large chain link to each end of a 5-in. (12.5-cm) length of chain. (Alternatively, ask your local home-improvement store to make a chain-link hanger.) Thread one chain link through a 1-in. (2.5-cm) pipe clamp (clip), then attach the clamp to the top of the birdhouse, following the instructions on page 132.

9 If you wish, paint the pipe clamp with dark brown exterior craft paint and let dry.

The dream of flying like a bird is as old as mankind itself, but airplanes have only been around for two centuries. Imagine that! With everyone wanting to fly through the air like a bird, I decided to create this unusual birdhouse airplane to accommodate those winged pilots.

airplane birdhouse

materials

Two 6ft x 5½ in. x ⅝ in. (180 x 19 x 2 cm) cedar dog-ear fence boards
48-in. (1.2-m) length of ¾ x ¾-in. (2 x 2-cm) square dowel
Waterproof premium glue
Wood putty
80-grit sandpaper
1-in. (25-mm) finish nails or galvanized wire nails
Twelve decorative 1-in. (25-mm) screws
Two 1¼ –in. (30-mm) screws
Four ½-in. (12-mm) screws
Six panel nails (these have ribs to hold better)
Two wood joiner nails
Red, dark blue, and metallic silver exterior spray paint
Two closet (wardrobe) rail sockets
Two drawer track rollers
Two picture hangers
24 in. (60 cm) #16 (1.6 mm) single jack zinc-plated chain
48-in. (1.2-m) length of 18-gauge (1-mm) wire
Basic tool kit (see page 126)
Needle-nose pliers

finished size

Approx. 14 x 6¾ x 5½ in. (35.5 x 17 x 14 cm)
Wing span: 23 in. (58.5 cm)

interior dimensions

Floor area: 5½ x 11 in. (14 x 28 cm)
Cavity depth: 4½ in. (11.5 cm)
Entrance hole to floor: 3 in. (7.5 cm)
Entrance hole: 1⅜ in. (35 mm) in diameter

cutting list

Sides: 13 x 5½ in. (33 x 14 cm)—cut 2, then mark the center of each short end and cut both corners off one end at 30° and both corners off the other end at 50° *
Top and bottom: 8¼ x 5½ in. (21 x 14 cm)—cut 2, beveling one short end at 30° and the other at 50°, with the bevels going in opposite directions
End tails: 4½ x 5½ in. (11.5 x 14 cm)—cut 2, beveling one short end at 50°
Front: 3⅛ x 5 ½ in. (8 x 14 cm)—cut 2, beveling one short end at 30°
Wings: 8 x 5½ in. (20 x 14 cm)—cut 2, then cut each one in half widthwise to give four wings measuring 2¾ x 8 in. (7 x 20 cm)
Propellers: 5½ x 5½ in. (14 x 14 cm)—cut 1, then cut two 5½ x ½-in. (14 x 1.2-cm) strips from this; cut one strip in half to give two 2½ x ½-in. (6 cm x 12 mm) strips

note

* Once you've made the initial cuts for the front and back pieces, it's easiest to cut the peaks by sandwiching them together and cutting through both layers at the same time. Reserve two of the small cut-off triangles for the rudder in step 12.

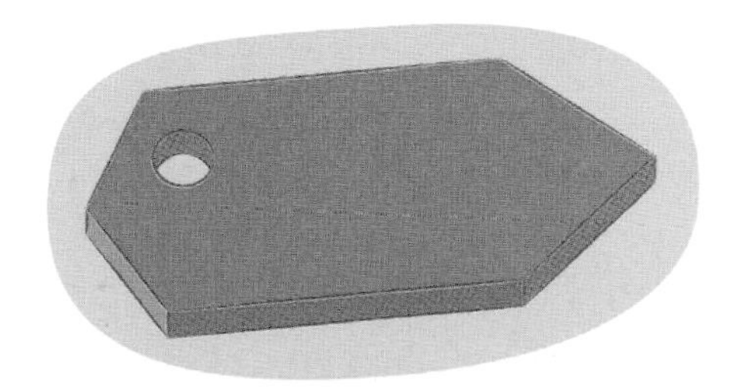

1 Cut a 1⅜-in. (35-mm) entrance hole in one side piece 2 in. (5 cm) from one long edge and 2½ in. (6 cm) from the 30° peak (see page 130).

Always dry fit your pieces to ensure the lengths match up on each edge. Give yourself a little leeway by making a longer cut, dry fit, then cut gradually so the pieces fit together perfectly.

2 Glue and nail the top and bottom pieces to the inner edges of the side piece with the entrance hole, as shown, with the 30° beveled edges next to the 30° peak.

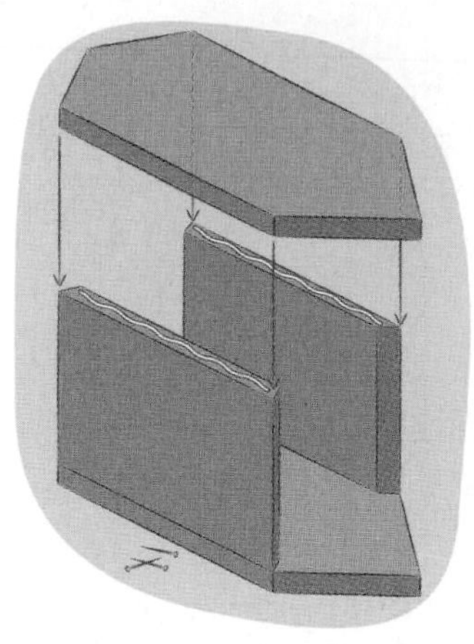

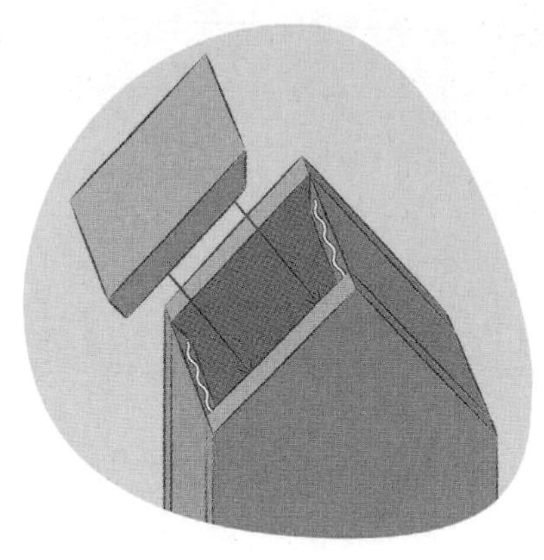

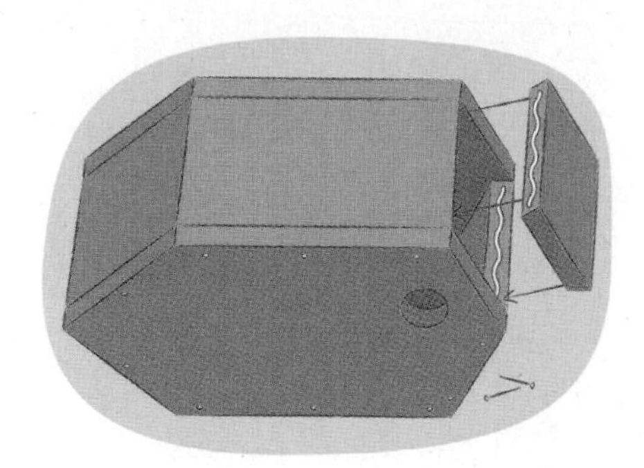

3 Apply glue to the edges of the top and bottom pieces and attach the other side piece, making sure the edges are flush and a good fit.

4 Apply glue to the 50° peak edges and fit the end tails, with the beveled edges aligning at the peak. Nail in place at an angle, making sure not to let the nails protrude.

5 Attach the front pieces to the 30° peak in the same way. Plug the entrance hole, paint the whole piece red, and let dry.

6 Attach the closet (wardrobe) rail socket over the hole, with the flange protruding outward. Screw in place. If you wish, add a second socket to the other side, so that both sides look the same.

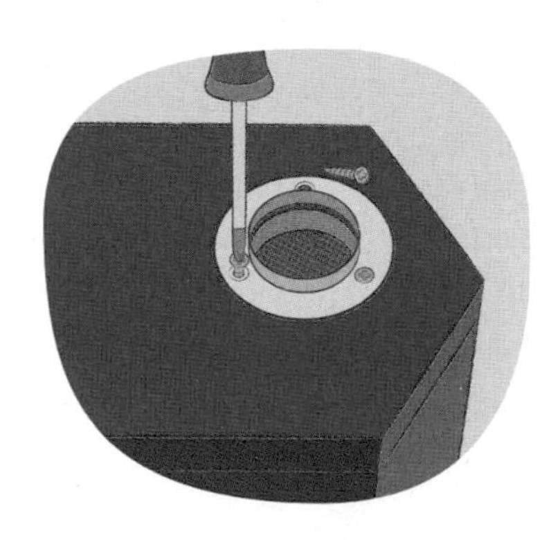

7 On two corners of each wing, cut off a ¼-in. (6-mm) right angle triangle, leaving the opposite short end straight cut. Cut four 4⅜-in. (11-cm) lengths of square dowel for the wing separators. Paint all the pieces dark blue and let dry.

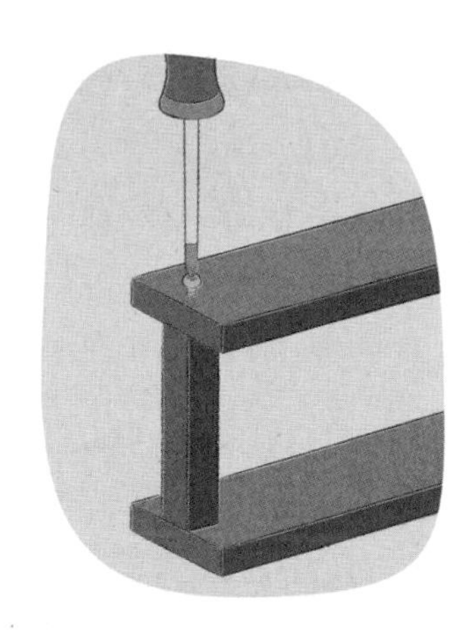

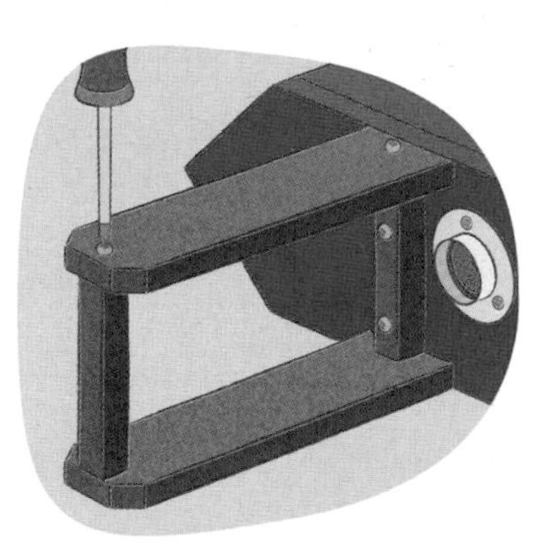

8 Center a wing separator on the short, straight-cut end of the wing. Drill a ⅛-in. (3-mm) pilot hole (see page 131) on the end of the wing and drive in a 1-in. (25-mm) decorative screw to join the wing and wing separator together. Attach a second wing piece on top.

9 Drill ⅛-in. (3-mm) pilot holes in the wing separator, then attach it to the side of the plane, near the entrance hole and flush with the top and bottom.

10 Attach a wing separator to the other end of the wing, in the same way as in step 8.

11 Repeat steps 8–10 to attach the second wing, using a square in order to line the wings up directly across from each other.

12 Cut a 9-in. (23-cm) length of square dowel for the rudder. Paint it dark blue, along with two of the small cut-off triangles reserved from cutting the sides, and let dry.

13 Place the rudder on the end tail, with the same amount protruding at each side, and drill a ⅛-in. (3-mm) pilot hole about 1½ in. (4 cm) from each end. Insert 1¼-in. (30-mm) decorative screws. Apply glue to each end of the rudder, then position the triangles with the points upward.

14 Assemble the propeller pieces as shown, using glue and wood joiner nails on both sides. Paint metallic silver on all sides and let dry.

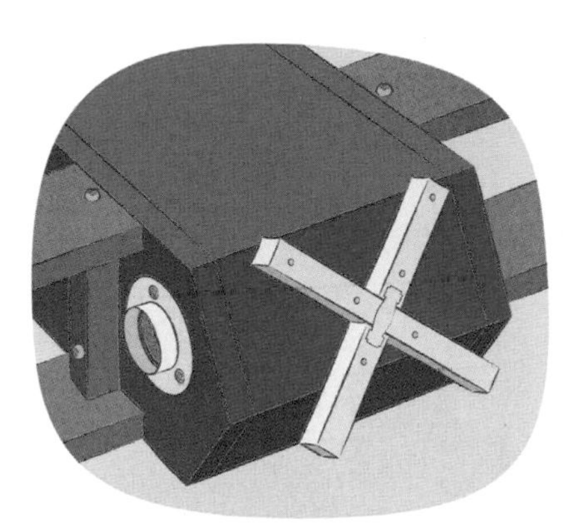

15 Glue the propeller to the front of the plane, then drill pilot holes with a 1⁄16-in. (1.5-mm) bit, 1½ in. (4 cm) apart and ¼ in. (6 mm) from the tip. Gently hammer in 1-in. (25-mm) panel nails.

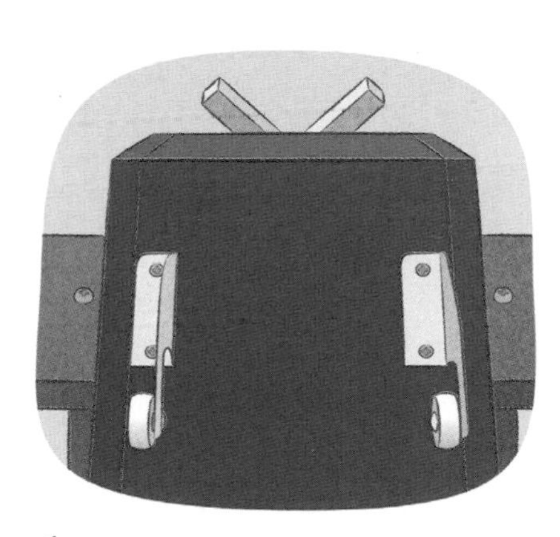

16 For the landing gear, attach two drawer track rollers to the bottom of the plane, next to the seam edges.

17 Using ½-in. (12-mm) screws, attach a picture hanger to each end of the top panel, ¼ in. (6 mm) from the edge. Cut a 24-in. (60-cm) length of single jack chain, open the first and last links with needle-nose pliers, feed the links through the framer hanger, and close them again.

18 Cut four 12-in. (30-cm) lengths of 18-gauge (1-mm) wire. Drill 1⁄16-in. (1.5-mm) holes through the top and bottom of both ends of the wings, 1½ in. (4 cm) from the screws on the wings. Pull the wires through the holes to form a diagonal cross, then coil the ends to tighten and secure the wires.

materials

Two 6ft x 7½ in. x ¾ in. (180 x 19 x 2 cm) cedar dog-ear fence boards
Waterproof premium glue
80-grit sandpaper
1-in. and 1¼-in. (25-mm and 30-mm) finish nails or galvanized wire nails
Two ¾-in. (20-mm) screws
Two 1-in. (2.5-cm) utility hinges
Weatherproof 30-minute clear silicone
Weathered fence pickets and/or driftwood pieces, 3–13 in. (7.5–33 cm) long
Green moss
Single bark piece, 9 in. (23 cm) long
5-in. (12.5-cm) length of 18-gauge (1-mm) wire
Basic tool kit (see page 126)
Jigsaw
Needle-nose pliers

finished size

Approx. 21 x 10 x 7½ in. (53 x 25 x 19 cm)

interior dimensions

Floor area: 6 x 7½ in. (15 x 19 cm)
Cavity depth: 14–18 in. (35.5 x 45.5 cm)
Entrance hole to floor: 12 in. (30 cm)
Entrance holes: 2 in. (5 cm) square

cutting list

Back: 21 x 7½ in. (53 x 19 cm)—cut 1
Front: 15 x 7½ in. (38 x 19 cm)—cut 1, beveling one short end at 30°
Sides: 15 x 7½ in. (38 x 19 cm)—cut 2, then cut across the board from the tip of one short end at a 30° angle
Floor: 6¼ x 7¼ in. (16 x 18.5 cm)—cut 1
Roof: 11¼ x 7 ½ in. (28.5 x 19 cm)—cut 1, beveling one short end at 30°

Along with the cute little cavity-nesting birds are the woodpeckers, easy to hear with their wood-drilling sound. They use man-made (or should I say "woman-made"?) bird boxes for nesting. Here's a cool, raw birdhouse ideally suited to them. Simple lines with natural elements of weathered wood, moss, and bark are ideal for our hammering feathered friends.

Woodpecker

woodpecker house

1 Cut the tips off the highest sides of the angled side pieces, ¾ in. (2 cm) down. This will create ventilation.

2 On the front panel, mark 11 in. (28 cm) from the bottom and 1½ in. (4 cm) from the top, centering your marks widthwise. Cut a 1½-in. (40-mm) hole with a hole saw, then use a jigsaw to enlarge the hole to 2 in. (5 cm) square. Alternatively, use a 2-in. (5-cm) hole saw to make a round hole. Sand the edges.

In this project we do not use any chemicals or paints, since woodpeckers are wood-drilling maniacs and could ingest them. Leave the wood raw and don't use putty to fill the holes.

3 Apply glue to the shortest edge of the side panel, then nail the front panel in place. Attach the other side panel to the front in the same way.

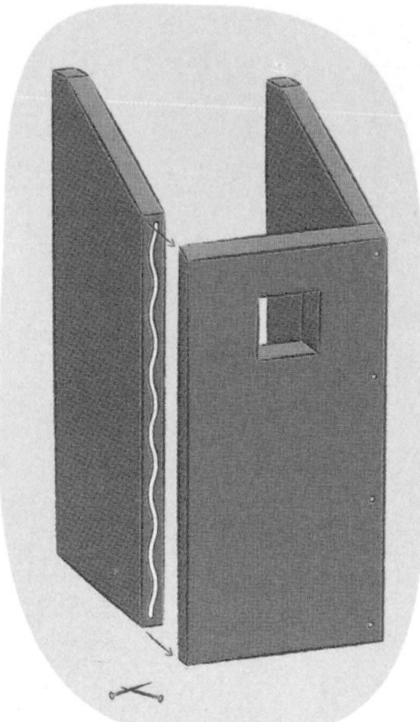

4 Dry fit the floor panel to ensure it fits between the sides and is flush to the edges, then glue and nail it in place.

5 Apply glue to the long side edges, then nail the back panel in place. Approx. 2½ in. (6 cm) will stick up above the sides.

Did you know that woodpeckers are farmers? Well...not really. But some species of these feathered geniuses gather acorns and stuff them into trees where they've drilled a hole purposely to store their food source. Not only is it for the nuts to get them through winter, but they find the acorns that have some larvae in them and that's an extra treat.

6 With the 30° beveled edge at the top, position the roof on top of the the side edges so that it touches the back. Drill pilot holes (see page 131) for two hinges 1 in. (2.5 cm) in from the side edges, then insert the screws. Then attach the other side of the hinges to the back panel in the same way.

Dry fit the hinges to ensure you have the correct placement and that they open in the correct direction.

7 For a rustic finish, attach weathered bits of fence or driftwood, about 3 to 13 in. (7.5 to 33 cm) in length, to the front, attaching them first with silicone and then with nails. Stick pieces of moss in and around the driftwood with silicone. Attach a piece of bark (make sure it's bug free!) to the roof in the same way, securing it with exterior screws, then cover the screw heads with moss.

8 Cut a 5-in. (12.5-cm) length of 18-gauge (1.2-mm) wire and attach it to the back panel with ¾-in. (20-mm) screws to form a hanging loop (see page 132). The woodpecker house is now ready to hang on a tree, preferably up high—12–20 feet (3.5–6 m) above the ground.

The length of the nails and screws will depend on the thickness of the pieces; use 1-in. (25-mm) nails for small pieces, 1¼-in. (30-mm) nails for thicker ones, and screws for the bark.

This is a fun project for the whole family to get involved in, and it will teach children all about the birds in the garden. What child wouldn't want to help make something pretty for birds to nest in and rear their young?

flower and ladybug birdhouse

materials

One 6 ft x 5½ in. x ½ in. (180 x 14 x 1.2 cm) dog-ear fence panel
Waterproof premium glue
1-in. (25-mm) finish nails or galvanized wire nails
Two 1¼-in. (30-mm) exterior screws
1-in. (25-mm) EMT two-hole pipe clamp (clip)
Two ¾-in. (20-mm) wood screws
Paintable wood-filler putty
80-grit sandpaper
Gray exterior wood primer paint
Pink oil-based exterior spray paint
Dark blue, bright blue, ivy, light green, dark green, lemon, black, and dark brown exterior craft paints
Ladybug stick pins
Glass door knob or drawer knob, approx. 1–1½ in. (2.5–4 cm) in diameter
1¾-in. (4.5-cm) length of round dowel, 5⁄16 in. (8 mm) in diameter
Water-based exterior varnish
Basic tool kit (see page 126)

finished size

Approx. 11 x 6¾ x 5½ in. (28 x 17 x 14 cm)

interior dimensions

Floor area: 4¼ x 5½ in. (10.75 x 14 cm)
Cavity depth: 8 in. (20 cm)
Entrance hole to floor: 6 in. (15 cm)
Entrance hole: 1¼ in. (32 mm) in diameter

cutting list

Front and back: 9 x 5½ in. (23 x 14 cm)—cut 2
Sides: 6¼ x 5½ in. (16 x 14 cm)—cut 2 (one side is reserved for the door)
Bottom roof: 5 x 5½ in. (20 x 14 cm)—cut 2
Top roof: 5¾ x 5½ in. (14.5 x 14 cm)—cut 2
Floor: 4¼ x 5½ in. (11 x 14 cm)—cut 1

1 Referring to the Basic Birdhouse on page 8, cut and shape the birdhouse pieces from dog-ear fence board. Cut a 1¼-in. (32-mm) entrance hole, 6½ in. (16.5 cm) from the bottom of the front panel and centered on the width.

2 Assemble the birdhouse and prepare it for painting, following steps 2–8 of the Basic Birdhouse on pages 8–9.

3 Attach a 1-in. (25-mm) EMT two-hole pipe clamp (clip) to the apex of the roof for a hanging loop (see page 132).

4 Prime the birdhouse (including the base) with gray exterior primer paint. Leave to dry. Go over the primer with any color of exterior paint you choose; I used bright pink. Don't worry if some of the primer color shows through a little in places—it looks cool and gives the birdhouse a slightly worn, weathered look. Paint the roof in dark brown exterior craft paint.

5 The painting technique used here uses two colors at a time on the same brush—dip half of the tip in one color and the other half in the second color. (Practice on white paper first.) First, use a ¾-in. (2-cm) brush and bright green and light green paint to make long stems around the base of the birdhouse. Then repeat the process, using a ½-in. (12-mm) brush and lemon yellow and light green paint, making fewer strokes this time. Finally, using a long, thin brush, paint longer lines in dark green and lemon yellow, making an odd number of strokes on each birdhouse panel.

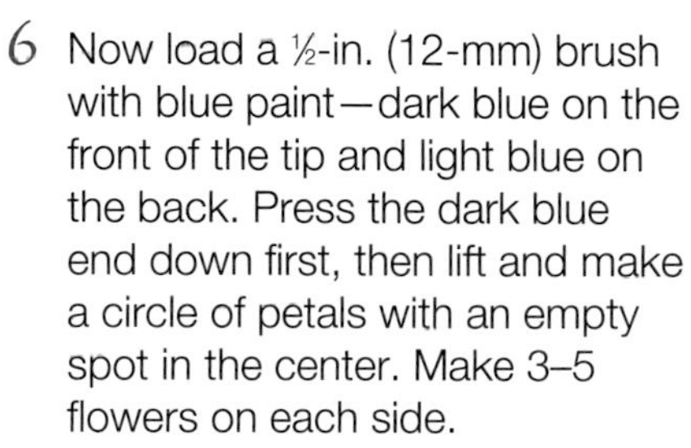

6 Now load a ½-in. (12-mm) brush with blue paint—dark blue on the front of the tip and light blue on the back. Press the dark blue end down first, then lift and make a circle of petals with an empty spot in the center. Make 3–5 flowers on each side.

7 Using a small, thin brush, fill in the flower centers with black paint. While it is still wet, pull lines of paint outward all around to create spiderlike centers. With a long slim brush, paint lemon yellow stripes all around the petals to give them more depth.

8 Using a 1/16-in. (1.5-mm) pin bit, drill a small pilot hole for the ladybug pins. With a 5/16-in. (8-mm) bit, make a small indention around the pin holes. Apply waterproof glue to the holes and place the pins in position. Using a a finish hammer, gently tap the pins into the wood. Attach 3–5 ladybugs to the front panel; add more to the other panels if you wish.

9 Using a 3/16-in. (5-mm) bit, drill a hole from the back of the door and attach the door knob. I offset the door knob to make it a little more interesting.

10 Drill a hole for the dowel perch and insert the perch into the hole (see step 9 on page 9). Paint the perch to match the flowers. Plug the entrance hole with wadded-up paper or tape, then varnish the exterior of the birdhouse (see page 127).

With lengths of chain dangling down, this birdhouse reminds me of an old barn with farming equipment attached and reminds me of spending time with my Gramps, rounding up the cattle. I used broken chain from a hanging basket—a great way of recycling junk! A natural knot in the wood provides the entrance hole; knots should be at least 1 to 1½ in. (2.5 to 4 cm) in diameter for birds to be able to enter the box and nest.

Pied Flycatcher

knot-hole birdhouse with chains

materials

One 6ft x 7½ in. x ½ in. (180 x 19 x 1.2 cm) dark wood dog-ear fence board with knots
One 6ft x 5½ in. x ½ in. (180 x 14 x 1.2 cm) dark wood dog-ear fence board
Waterproof premium glue
1-in. (25-mm) finish nails or galvanized wire nails
80-grit sandpaper
Two 1¼-in. (30-mm) exterior screws
Two 2½-in. (60-mm) exterior screw
Five ¾-in. (20-mm) black decorative screws
40 in. (1 m) black small-link chain
Water-based exterior varnish
Basic tool kit (see page 126)

finished size

Approx. 14 x 9 x 9 in. (35.5 x 23 x 23 cm)

interior dimensions

Floor area: 6¼ x 5½ in. (16 x 14 cm)
Cavity depth: 12 in. (30 cm)
Entrance hole to floor: 8 in. (20 cm)
Entrance hole: 1–1½ in. (25–40 mm) in diameter

cutting list

Front and back panels: 13¼ x 7½ in. (33.5 x 19 cm)—cut 2 (one with a knot hole, one without)
Right side panel: 8½ x 5½ in. (21.5 x 14 cm)—cut 1 for door
Left side panel: 13¼ x 5½ in. (33.5 x 14 cm)—cut 1
Floor: 6¼ x 5½ in. (16 x 14 cm)—cut 1
Flat roof: 7½ x 4½ in. (19 x 11.5 cm)—cut 1
Pitched roof: 7 x 7½ in. (17.5 x 19 cm)—cut 1
Door façade: 7 x 3 in. (17.5 x 7.5 cm)—cut 1 with a knot hole in the center
Awning: 2½ x 4½ in. (6.5 x 11.5 cm)—cut 1

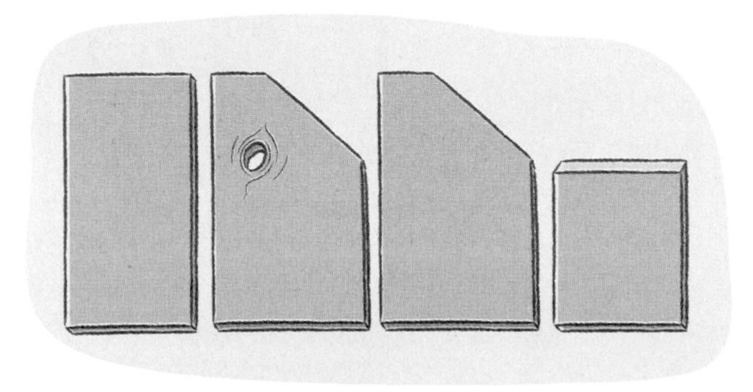

1 In the wood for the front panel, find a knot that's in the right place, slightly off center. Cut the front panel, cutting the base of the panel 8½ in. (21.5 cm) below the knot. Carefully cut out the knot if necessary; ideally, you want a piece of board where the knot has already gone or is loose, so that you don't have to cut it, but this may not be possible. Cut the back panel from a section of the board that does not have a knot. Using a speed square, mark 3 in. (7.5 cm) along the top of the front and back panels and cut from here to the side at a 45° angle. Reserve the off-cuts to use later. Bevel one short edge of the right side panel (the door) at 45°; do not bevel the left side panel.

2 Assemble the body of the birdhouse, following steps 2–4 of the Basic Birdhouse on page 8.

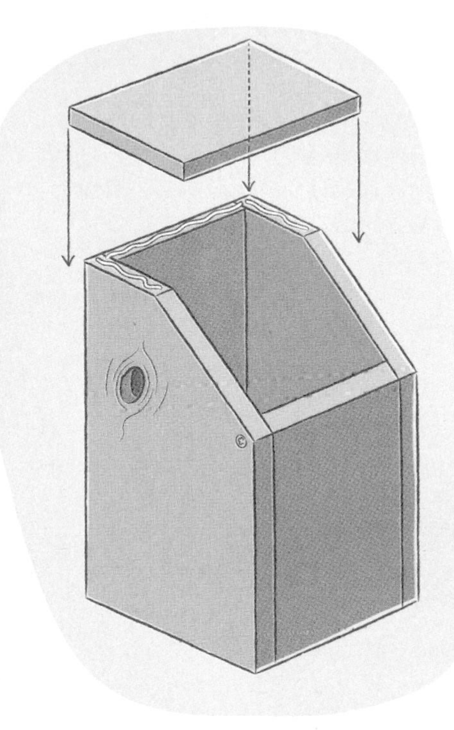

3 Glue and nail the flat roof in place, flush with the edge of the back and left side panels and overhanging the door by 1½ in. (4 cm) and the front panel by ½ in. (1 cm).

4 Bevel one short end of the pitched roof panel at 45°. Glue and nail the pitched roof panel under the flat roof, with the beveled edge uppermost.

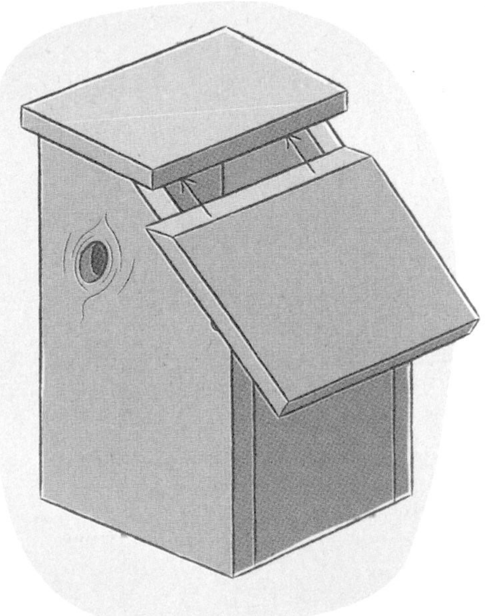

5 Place the off-cuts from the front and back panels together to form a square. Glue and nail them to the door to form the door pull. If you wish, drill in 1¼-in. (30-mm) screws from the inside of the door through to the door pull, for extra strength.

6 Remove the door. Using a ⅛-in. (3-mm) bit, drill pilot holes in the panel opposite the door, then drill in 2½-in. (60-mm) exterior screws so that you can mount the birdhouse on a tree or fence post.

7 Cut a ¼-in. (6-mm) sliver off each short end of the door façade. Glue and nail the door façade to the front of the birdhouse, 1 in. (2.5 cm) from the right-hand side and flush with the bottom. Using a 1⁄16-in. (1.5-mm) bit, drill a tiny hole in each end of the slivers of wood. Glue one sliver above and one below the knot in the door façade and gently hammer in 1-in. (25-mm) nails to attach.

8 Bevel one long edge of the awning at 45°. Glue and nail it in place on the front panel, above the entrance hole.

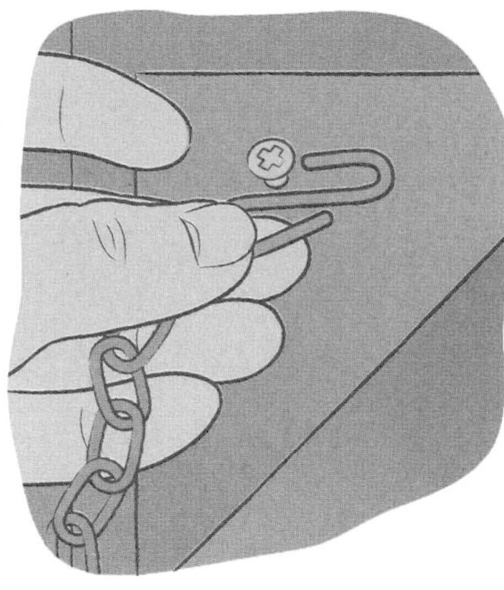

9 Decide where you want to attach the decorative chains. Using a ⅛-in. (3-mm) bit, drill a pilot hole at each attachment point. Drive a ¾-in. (20-mm) decorative black screw partway in at each point, open a chain link, wrap it around the screw, then close the link again. Tighten the screw to hold the chain in place.

10 Protect the entrance hole with wadded-up paper or tape, then varnish the exterior of the birdhouse (see page 127).

11 Drill a 5⁄16-in. (8-mm) hole in the center of the floor for drainage. Drill 5⁄16-in. (8-mm) ventilation holes in the door, just under the eaves, slanting them slightly upward to prevent rainwater from seeping in.

An attractive novelty birdhouse that can be painted to look like your favorite bird, this sits high on galvanized fence posting and has ventilation holes on the back end and the front, just under the beak.

bird's eye birdhouse

materials

One 6 ft x 7½ in. x 2½ in. (180 x 19 x 1.2 cm) dog-ear fence board
One 6 ft x 5½ in. x 2½ in. (180 x 14 x 1.2 cm) dog-ear fence board
10 x 4 x ½ in. (25 x 10 x 1.2 cm) lumber (timber)
1¾-in. (4.5-cm) length of 5⁄16-in. (8-mm) round dowel
80-grit sandpaper
Waterproof premium glue
1-in. (25-mm) finish nails or galvanized wire nails
Medium blue oil-based exterior spray paint
Orange, light blue, light grey, and black exterior craft paint
Nine 1¼-in. (30-mm) exterior screws
Water-based exterior varnish
Galvanized hollow steel fence post
Four ¼-in. (6-mm) nuts
1-in. (2.5-cm) HS hub connector (from electrical section of home store)
1 x 6-in. (2.5 x 15-cm) galvanized nipple
Four 1⅝-in. (40-mm) exterior screws
Two 4-foot (1.2-m) rebar (reinforcing bars), ½ in. (12 mm) in diameter
Basic tool kit (see page 126)

finished size

Approx. 10 x 20 x 7¾ in. (25 x 50 x 19.5 cm)

interior dimensions

Floor area: 13 x 5½ in. (33 x 14 cm)
Cavity depth: 6¾ in. (17 cm)
Entrance hole to floor: 6 in. (15 cm)
Entrance hole: 1½ in. (40 mm) in diameter

cutting list

Front and back: 14 x 2¾ x 8 x 2¾ x 10¾ in. (35.5 x 7 x 20 x 7 x 27.3 cm)—cut 2 from 7½-in. (19-cm) fence board (see step 1)
Top panel: 13 x 5½ in. (33 x 14 cm)—cut 1 from straight end of 5½-in. (14-cm) fence board
Beak: 3½ x 5½ in. (9 x 14 cm)—cut 1 from dog-ear end of 5½-in. (14-cm) fence board
Butt: 9¾ x 5½ in. (24.7 x 14 cm)—cut 1 from 5½-in. (14-cm) fence board
Floor panel: 2½ x 5½ in. (6.3 x 14 cm)—cut 1 from 5½-in. (14-cm) fence board
Breast panel: 7¼ x 5½ in. (18.5 cm)—cut 1 from 5½-in. (14-cm) fence board
Base panel: 6 x 5½ in. (15 x 14 cm)—cut 1 from 7½-in. (19-cm) fence board
Roof panel 1: 16 x 7½ in. (40.5 x 19 cm)—cut 1 from straight end of 7½-in. (19-cm) fence board
Roof panel 2: 5 x 7½ in. (12.5 x 19 cm)—cut 1 from dog-ear end of 7½-in. (19-cm) fence board
Wing: 10 x 4 in. (25 x 10 cm)—cut 1 from lumber (timber)

1 Cut the front panel from the dog-ear end of 7½-in. (19-cm) fence board. This may look complicated, but you simply measure 14 in. (35.5 cm) along one edge from the base of the dog ear and make a straight cut across the board, and then make a 45° cut on each end of the board, as shown in the diagram. Then place the front panel on the remaining board and draw around it, and cut out the back panel; this way, the two panels are identical.

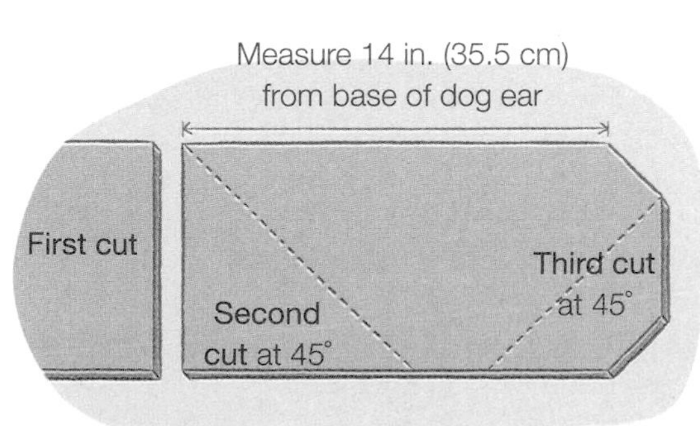

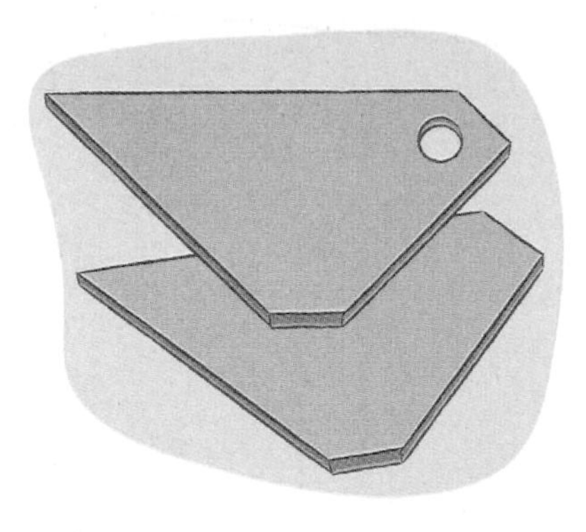

2 Cut a 1½-in. (40-mm) entrance hole in the front panel 1½ in. (4 cm) down from the top edge and 1½ in. (4 cm) from the point.

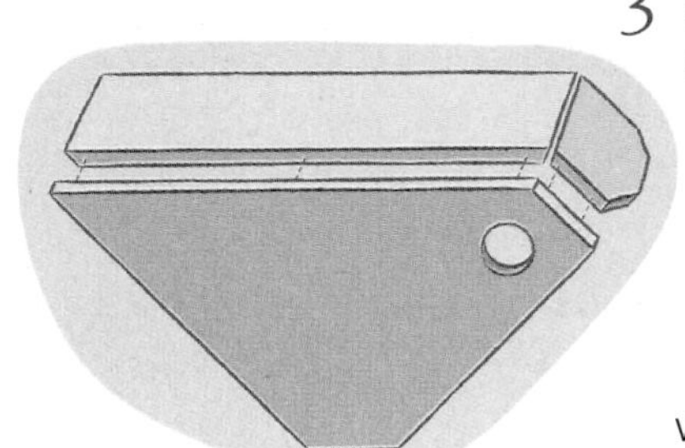

3 Bevel one short side of the top panel at 45° and the other short side at 22.5°. Glue and nail the front panel to the top panel edge, with the 22.5° bevel of the top panel at the entrance-hole end and the 45° bevel at the tail end. Glue and nail the back panel to the top panel in the same way. Bevel one long side of the beak at 22.5°, then glue and nail the beveled end of the beak to the end of the top panel.

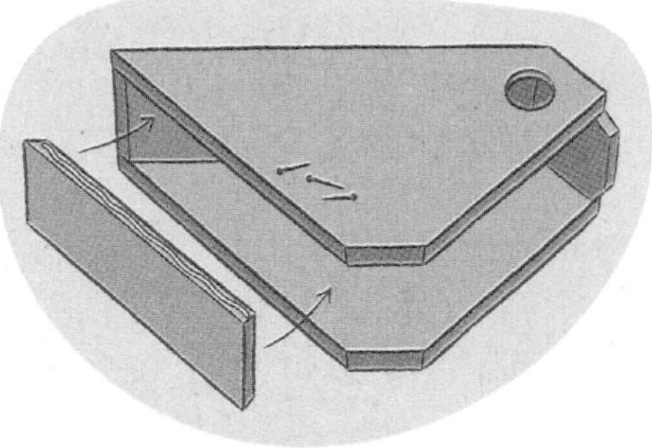

4 Bevel one short side of the butt at 45° and the other at 22.5°. Glue and nail the sides and 45° beveled end of the butt to the other end of the top panel, leaving the 22.5° end of the butt without glue.

5 Bevel both long sides of the floor panel at 22.5°. Dry fit the floor in place, to ensure that your cuts are accurate. Using a ⅛-in. (3-mm) bit, drill pilot holes in each side for 1¼-in. (30-mm) screws. In this project, the floor is unscrewed to clean out the birdhouse.

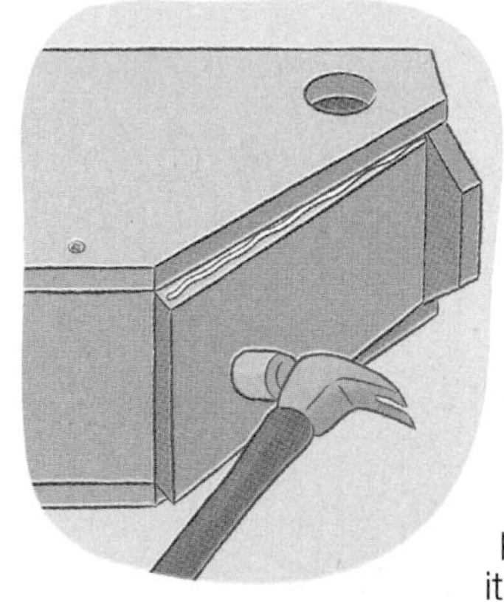

6 Bevel one short side of the breast panel at 22.5°. Glue and nail the straight-cut end and the sides of the breast in place, leaving the beveled end free of glue. You may have to tap the piece gently with a hammer to fit it in place.

7 Using a 3⁄16-in. (5-mm) bit, drill four holes for drainage in the floor. Using the pilot holes that you drilled in step 5, drill in 1¼-in. (30-mm) screws to attach the floor panel between the breast and butt panels.

8 Using a 5⁄16-in. (8-mm) bit, drill ventilation holes at a downward angle on the butt and breast panels.

9 Using a ⅛-in. (3-mm) bit, drill a pilot hole or a 1¼-in. (30-mm) screw in the middle of the base panel. This will be where the hub connector is attached once the birdhouse is complete.

10 On roof panel 1, bevel one short end at 22.5°. On roof panel 2, bevel the end opposite the dog ear at 22.5°. Cut the un-beveled short end of roof panel 1 at 45° on each side, leaving 1½ in. (4 cm) straight cut in the center. Glue and nail roof panels 1 and 2 to the top panel along their beveled edges, positioning them flush with the back panel and overhanging the front by 2 in. (5 cm). This will provide an awning to protect the entrance hole.

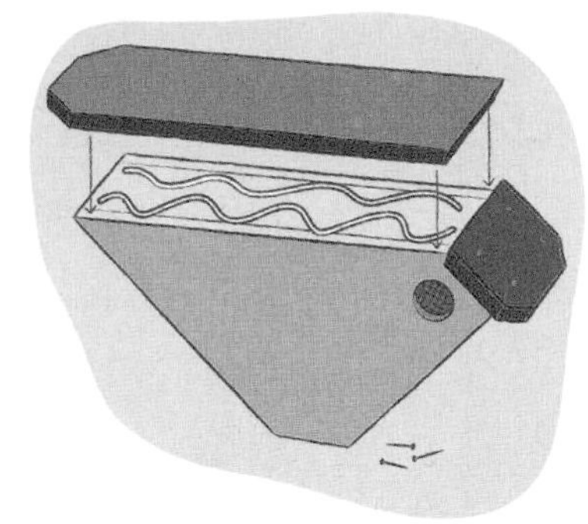

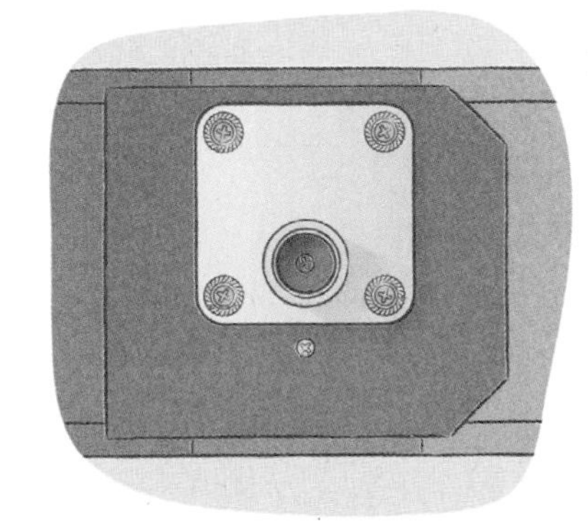

11 Make a 45° miter cut 1½ in. (4 cm) from the each edge on the bottom of the wing. Using a ⅛-in. (3-mm) bit, drill three pilot holes as shown. Glue the wing in place on the front of the bird, then drill in 1¼-in. (30-mm) exterior screws. Using a 5⁄16-in. (8-mm) bit, drill a hole for the dowel perch (see step 9 on page 9) and Insert the perch.

12 Prepare the birdhouse for painting (see step 10 on page 9), then paint it bright blue. Paint the base panel, too.

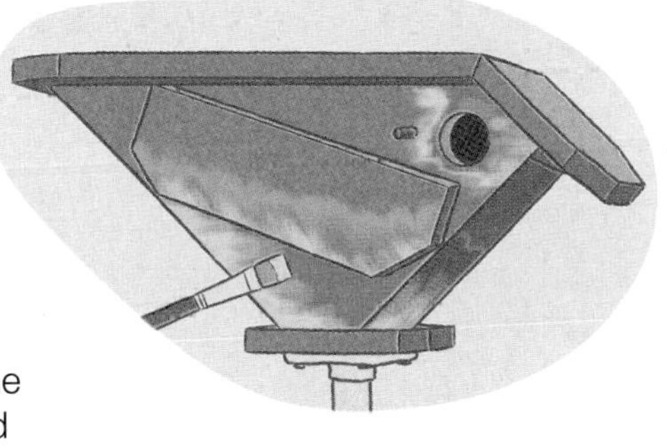

13 When the paint is dry, attach the base panel to the bottom of the birdhouse with a 1¼-in. (30-mm) screw. Drill in two more screws, one on either side of the central one. Attach a 1-in (2.5-cm) hub connector to the bottom of the floor, with the hole centered over the painted blue screw, using four ¼-in. (6-mm) nuts to ensure that the screws do not protrude through to the inside of birdhouse. Attach the nipple pipe.

14 Prepare the birdhouse for painting (see step 10 on page 9). I used colors similar to those on an Eastern Bluebird, applying orange exterior craft paint to the side, breast, butt, and around the eye area, and then blended in touches of light grey and light blue here and there.

15 Then I added black around the eye area, painted the beak solid black, and dry brushed black streaks around the bottom of the wing. If you wish, you can paint the back panel in the same way. Leave to dry. Plug the entrance hole with wadded-up paper or tape, then varnish the exterior of the birdhouse (see page 127).

16 Mount the birdhouse on a hollow galvanized fence pipe, then plant the post in your garden over 4-ft rebars (see page 135).

With magnets to hold the metal clean-out doors in place and coat hooks for landing perches, this is the ultimate triple-decker birdhouse. Any of the small cavity-nesting birds—wrens or chickadees, for example—will love it.

crooked birdhouse

materials

Two 6ft x 5½ in. x ⅝ in. (180 x 14 x 2 cm) cedar dog-ear fence boards
Waterproof premium glue
Wood putty
80-grit sandpaper
1-in. (25-mm) finish nails or galvanized wire nails
1¼-in. (30-mm) screws
Four ¾-in. (20-mm) screws
Three 3½ x 3¾ x ⅛ in. (9 cm x 9.5 cm x 3 mm) metal electrical covers
Six 11⁄16-in. (17-mm) magnets
Weatherproof 30-minute clear silicone
Gray and white exterior spray paint
D-ring picture hook, 2½ x 1 in. (6 x 2.5 cm)
11-in. (28-cm) length of 11-gauge (2.3-mm) chain
Three metal closet (wardrobe) rail sockets
Three single-prong coat hooks
Light sage green craft paint
Basic tool kit (see page 126)
Jigsaw
Needle-nose pliers

finished size

Approx. 27½ x 14 x 5½ in. (70 x 35.5 x 14 cm)

interior dimensions

Floor area: 10 in. x 5 in. x 4¼ in. (25 x 12.5 x 11 cm)
Cavity depth: 9 in. (23 cm)
Entrance hole to floor: 5 in. (12.5 cm)
Entrance hole: 1⅛ in. (28 mm) in diameter

cutting list

Fronts and backs: right-angle triangles measuring 11 x 7¾ in. (28 x 19.5 cm) —cut 6 *
Bases: 12¼ x 5½ in. (31 x 14 cm)—cut 3, beveling each short end at 45°, with the bevels going in opposite directions
Roof piece 1: 9¼ x 5½ in. (23.5 x 14 cm)—cut 3, beveling one short end of each at 45°
Roof piece 2: 9¾ x 5½ in. (24.5 x 14 cm)—cut 3, beveling one short end of each at 45°

note

* Start at the end of the fence board at a 45° angle and cut one triangle, then use that as a template to cut the rest.

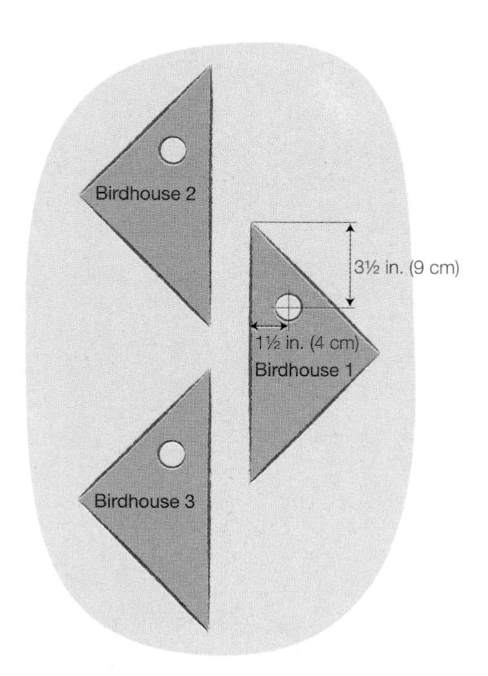

1 Cut 1⅛-in. (28-mm) entrance holes (see page 130) in the three front triangles, as shown, positioning each one 3½ in. (9 cm) from what will be the highest point of each birdhouse and 1½ in. (4 cm) in from the longest side.

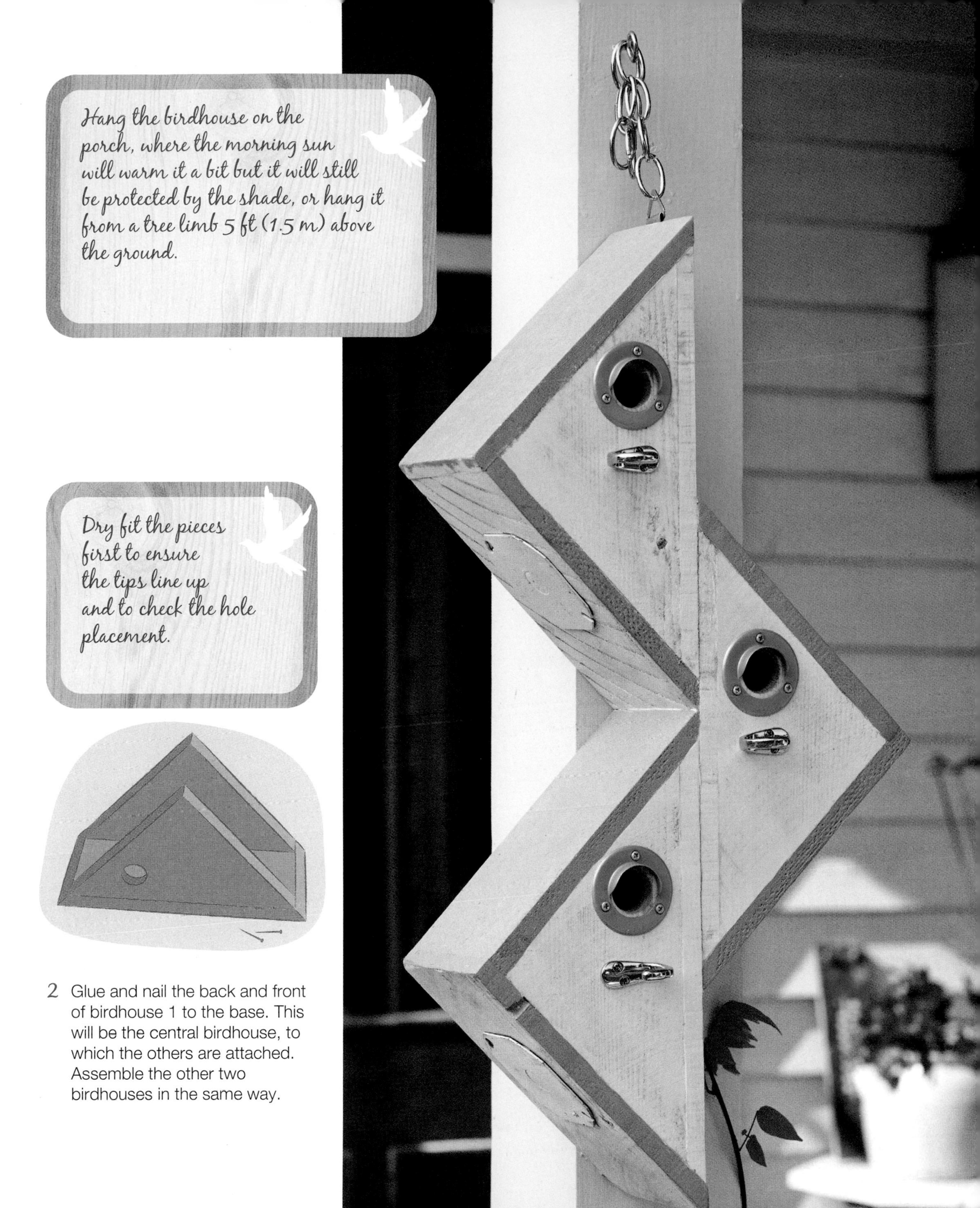

Hang the birdhouse on the porch, where the morning sun will warm it a bit but it will still be protected by the shade, or hang it from a tree limb 5 ft (1.5 m) above the ground.

Dry fit the pieces first to ensure the tips line up and to check the hole placement.

2 Glue and nail the back and front of birdhouse 1 to the base. This will be the central birdhouse, to which the others are attached. Assemble the other two birdhouses in the same way.

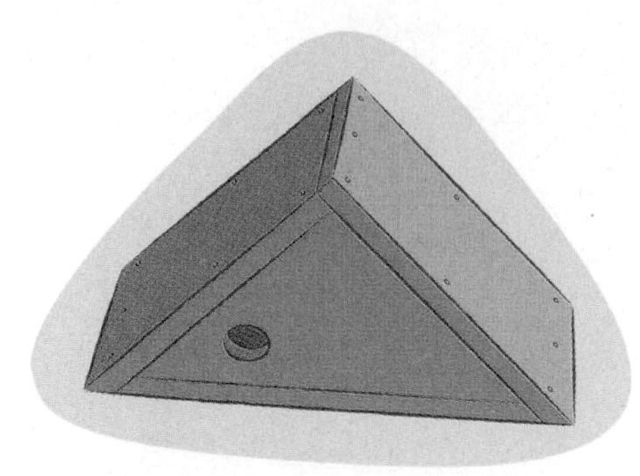

3 Glue and nail roof piece 1 in place on birdhouse 1, with the beveled edge flat on the table, then add roof piece 2 to complete the birdhouse.

4 Place birdhouse 1 on your work surface, with the hole facing up. Place birdhouse 2 on top, long edge to long edge, positioning it 7 in. (19 cm) up from the bottom of the birdhouse 1. Attach roof panel 1 to the *left-hand* short edge, then glue and nail the second birdhouse to birdhouse 1. Using a ⅛-in. (3-mm) bit, drill pilot holes (see page 131) and drive in 1¼-in. (30-mm) screws.

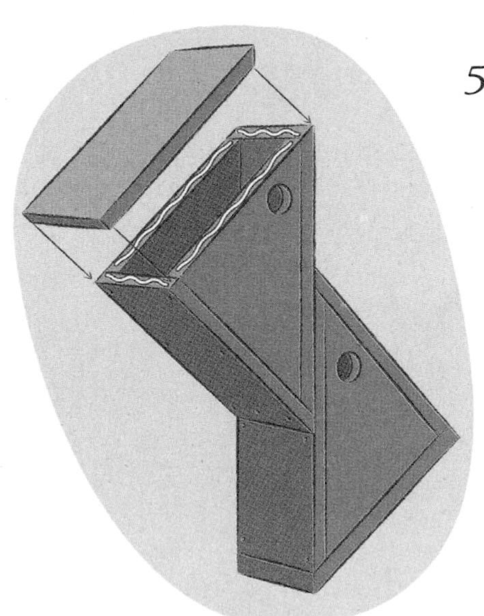

5 Now glue and nail the remaining roof panel in place on birdhouse 2.

6 Attach roof panel 1 to the *right-hand* short edge of birdhouse 3. Place birdhouse 3 on top of birdhouse 1, long edge to long edge, butting it up against birdhouse 2. Glue, nail, and screw the houses together, as in step 4, then add the final roof panel.

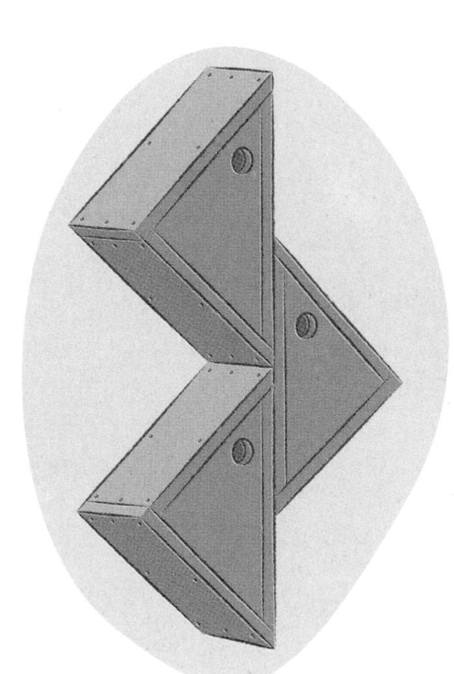

7 Now you're ready to cut the holes for the clean-out doors. On the bottom of each birdhouse, draw around a metal electrical plate cover with a pencil. Then, on diagonally opposite sides of the plate cover markings, draw around a magnet ⅛ in. (3 mm) away from the inside line to mark where the magnets will go.

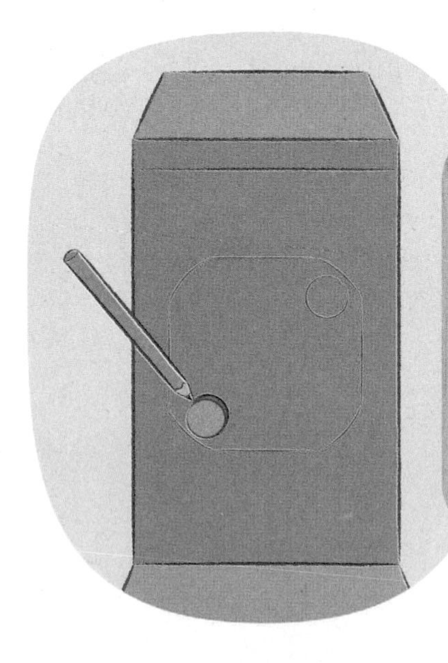

Make sure you mark the clean-out doors on the bottom of each birdhouse, not on the back panels.

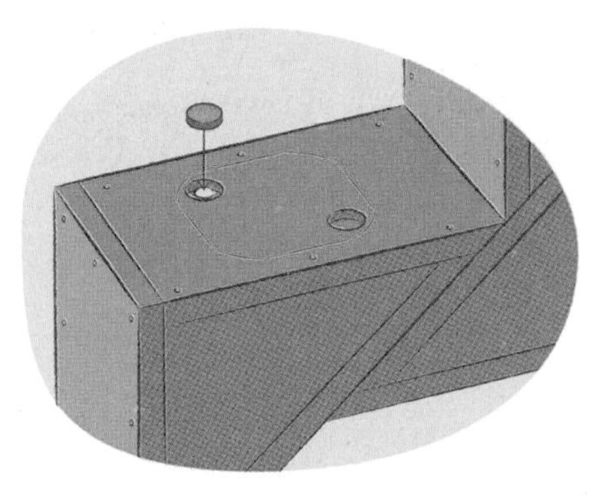

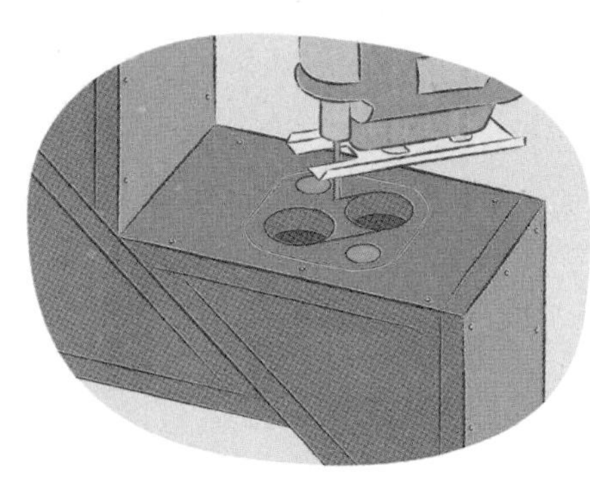

8 On the bottom of each box, using an 11⁄16-in. (17-mm) bit, drill two gullies so that the magnets will sit flush. **Do not drill all the way through**. Apply a drop of silicone, insert the magnets, and hammer them flush.

9 Using a 1½-in. (4-cm) hole saw, cut two holes ¼ in. (6 mm) away from each other and ¼ in. (6 mm) inside the pencil marks indicating where the plate cover will lie. Cut out the remaining wood between the two holes with a jigsaw to create a clear-out hole 3¼ in. (8 cm) long and 1½ in. (4 cm) wide. Cover the hole with the plate cover.

10 Fill the entrance holes with wadded paper or tape. Paint the entire birdhouse gray, including the plate covers, let dry, and then paint white over the gray.

11 Using pliers, open up the last link in the chain. Hook it through the D-ring of the picture hook, then close the link again. Drill small pilot holes in the base of the top birdhouse for the screws in the picture hanger, then attach the picture hook with ¾-in. (20-mm) screws.

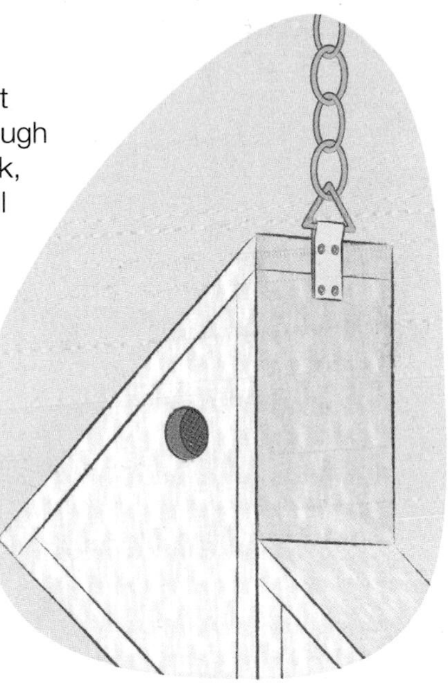

12 Using a ⅛-in. (3-mm) bit, drill pilot holes around the entrance hole for the closet (wardrobe) rail sockets, and below the holes for the coat hooks, placing the coat hooks sideways ½ in. (12 mm) below the entrance holes. Then screw in the pieces using the screws provided in the pack.

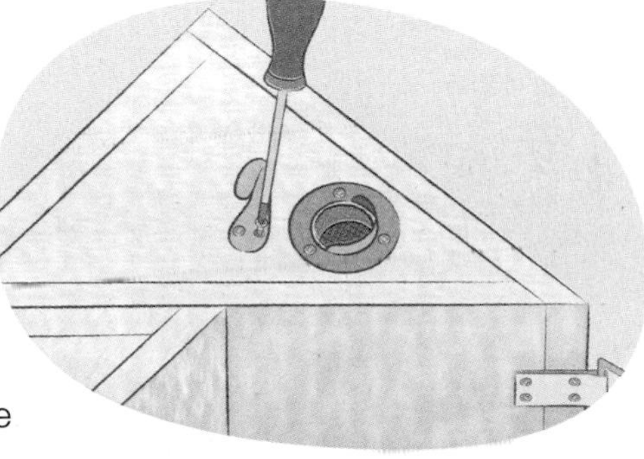

13 Paint the tops and front edges a light sage green or color of your choice.

chapter 2

bird feeders

It's always fun to watch birds while relaxing with a cup of tea in the morning, reading the local newspaper. Here are many attractive bird feeders made from recycled items such as teacups, wine bottles, and wood. You'll also find a basket to hold nesting materials and a swing for birds to perch upon.

House Sparrow

Once the birds discover this hot pink feeder filled with delightful delicacies such as sunflower seeds, peanuts, and bird seeds, they'll be flocking to your garden in droves.

swing bench bird feeder

materials

One 6ft x 5½ in. x ⅝ in. (180 x 14 x 1.5 cm) cedar dog-ear fence board
Waterproof premium glue
Wood putty
80-grit sandpaper
1-in. (25-mm) finish nails or galvanized wire nails
1-in. (25-mm) steel panel board nails
4 x 8 in. (10 x 20 cm) galvanized ⅛-in. (3-mm) metal mesh
Four #4 (2½-in./60-mm) zinc-plated steel screw eyes
60 in. (150 cm) #16 (1.6 mm) single jack zinc-plated chain
Basic tool kit (see page 126)
Needle-nose pliers

optional items for decoration

18-gauge (1-mm) rebar tie wire
Black furniture tacks
Flower stamp
Pink exterior spray paint
White, green, and yellow craft paint

finished size

Approx. 10½ x 10 x 6 in. (26.5 x 25 x 15 cm)
Trough: 1¾ x 4 x 6¾ in. (4.5 x 10 x 17 cm)

cutting list

Chair and feeder trough: 8 x 5½ in. (20 x 14 cm)—cut 3 (reserve 1 for step 2)
Arms: 6 x 5½ in. (15 x 14 cm)—cut 1

1 Cut two of the 8 x 5½-in. (20 x 14-cm) pieces at a 45° angle on both sides to create a peak. (Reserve the cut-off right-angle triangles for the feet—see step 5.) With the straight short edge against the fence of the chop saw, slice off 1½ in. (4 cm) on each long side of each panel, leaving the center sections 2 in. (5 cm) wide. (You'll need to flip the panel over before cutting the second side.) These will be used to make the backrest of the chair.

2 Cut the reserved 8 x 5½-in. (20 x 14-cm) piece lengthwise into three sections 1½ in. (4 cm) wide x 8 in. (20 cm) long. Then cut one of these three sections in half widthwise to give two 3½ x 1½-in. (9 x 4-cm) pieces. These pieces will be used to make the feeder trough.

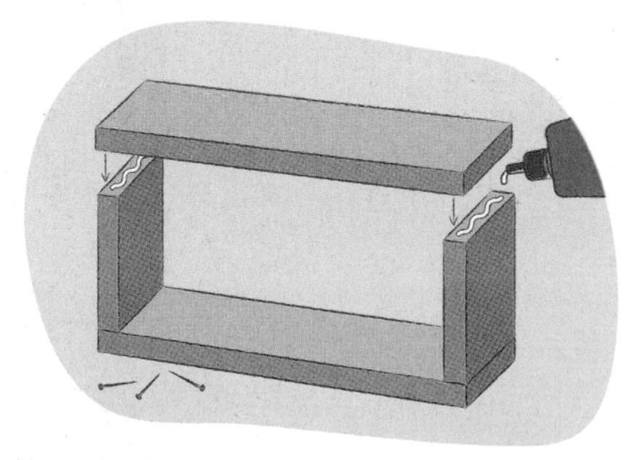

3 Glue and nail the four feeder trough pieces together to form a rectangular frame. Note that the long pieces need to be glued to the top of the short pieces to create a frame the right size for the mesh.

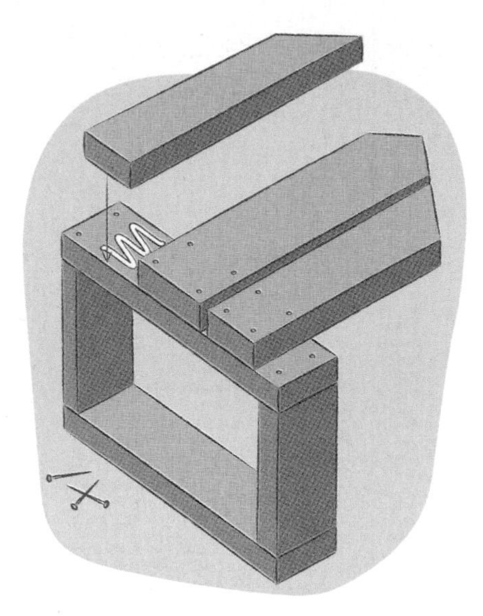

4 Now attach the angled pieces from step 1 to the feeder trough to create the backrest of the chair. Find the center of one long edge of the trough, then glue and nail a 2-in. (5-cm) picket in place. Then add a 1½-in. (4-cm) picket on either side, leaving a gap of about ¼ in. (6 mm) in between.

5 Next, add the feet to the feeder trough. Glue and nail the right-angle triangles from step 1 to each short side of the trough, setting the first two flush with the top and back edges of the trough. The triangles at the front of the trough will overhang the edge slightly, leaving room for the front façade.

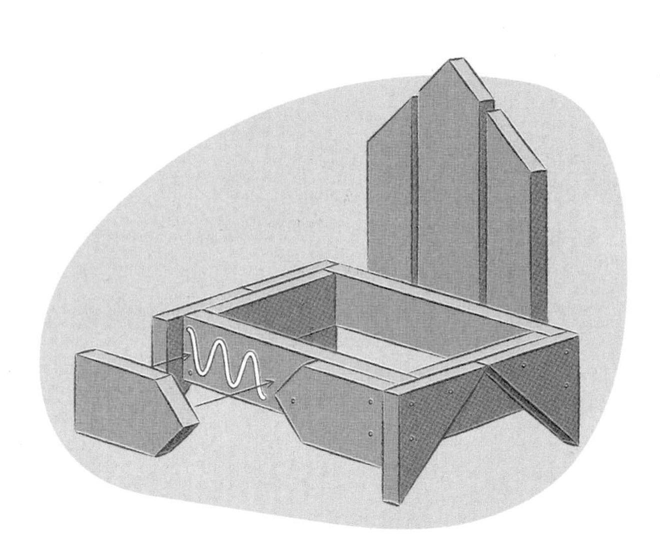

6 Take the remaining 2-in. (5-cm) picket, and cut the straight short end at a 45° angle on both sides to mirror the first peak. Cut the piece in half widthwise. Place each half on the center front of the trough, with the "arrowhead" pointing inward and the top edge aligning with the top edge of the trough, and glue and nail them in place to create the façade.

7 Now attach the last two pieces of the backrest—the 1½-in. (4-cm) pickets left over from step 1. Lay one piece against the back of the trough, aligning the pointed tip with the side picket so that you get a continuous slope. Mark where the bottom of the leg sits with a pencil, cut off the excess, and discard. Use the cut piece as a template to cut the remaining picket to size. Glue and nail them in place, aligning them with the outer edges of the legs.

8 From the piece that you cut for the arms, cut two 1½ x 6-in. (4 x 15-cm) pieces. Discard the leftovers. Then cut ¼ in. (6 mm) off at a 45° angle on three corners of each arm. Line up the end that only has one corner cut, flush with the back and inner trough edges. Glue and nail in place.

9 If you wish, lightly fill the nail holes with wood putty and sand lightly. Paint the feeder in your chosen color (I used pink) and let dry.

10 Turn the chair upside down, with the backrest against the edge of your table to support the chair while you hammer in the panel nails. Place wire mesh on the bottom of the feeder trough and cut to size. Using a 1⁄16-in. (1.5-mm) bit, drill three pilot holes (see page 131) on each side for the panel nails, spacing them evenly.

Make sure the nail heads are in the center of a mesh hole in order to trap the mesh on each side securely.

11 Using a 1⁄16-in. (1.5-mm) bit, drill a pilot hole in the center of each end of the arms about 1 in. (2.5 cm) back from the edge. Insert a #4 (2½-in./60 mm) screw eye into each one. Cut the chain into two 30-in. (75-cm) lengths. Using needle-nose pliers, attach one end of each one to the screw eyes on opposite corners of the chair, following the instructions on page 133. Use a single chain link to join the two chains together in the center for hanging.

12 If you wish, decorate the feeder. I used black furniture tacks and 18-gauge (1-mm) rebar tie wire twisted into spiral designs, which I attached by bending the ends of the wire over and hammering them into the wood (I drilled small pilot holes first). I also applied small flowers by dipping a flower-shaped stamp into white craft paint, then added green leaves and yellow flower centers by hand.

Never paint the internal areas of the feeder, as this could be very harmful to the birds.

hip bird feeder

This retro-style bird feeder brings back memories of beehive hair-dos, Beatles music, and simple living and will provide a groovy dining area for a variety of birds.

1 Cut the front and back pieces at a 30° angle on each side to form a peak. (Reserve the four cut-off triangles to decorate the front and back in step 9.) Cut a 1½-in. (40-mm) entrance hole in each piece, 1½ in. (4 cm) down from the center of the peak (see page 130).

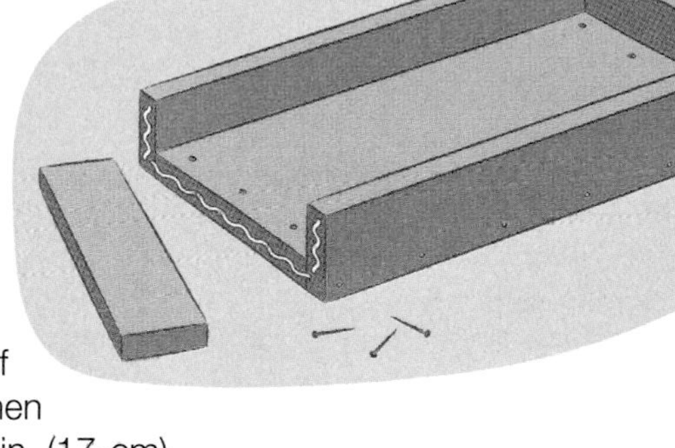

2 To assemble the feeder trough, glue and nail an 11½ x 2-in. (29 x 4-cm) furring strip to each long side of the feeder base. Then glue and nail a 6¾-in. (17-cm) furring strip to each short side.

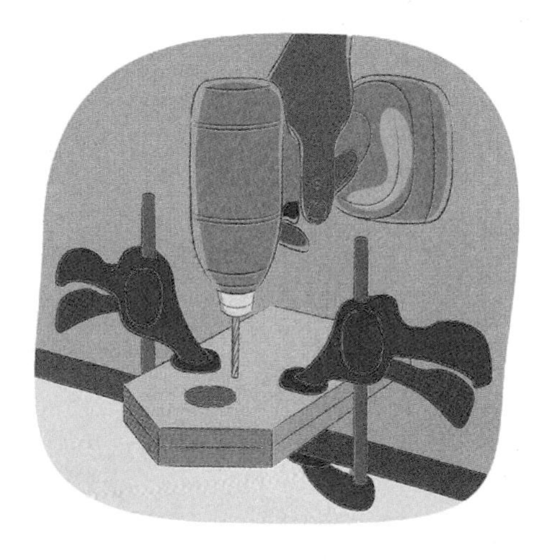

3 Clamp the front and back together on a steady surface. Drill a ⅜-in. (1-cm) hole 1 in. (2.5 cm) down from the bottom of the entrance hole, drilling all the way through both pieces. (A dowel will be inserted here as an internal perch, once the feeder has been assembled.)

materials

One 6 ft x 5½ in. x ⅝ in. (180 x 14 x 1.5 cm) cedar dog-ear fence board
One 8 ft x 2 in. x 1 in. (250 x 5 x 2.5 cm) furring strip board
12 in. (30 cm) of 1⅛ x 1⅛-in. (27 x 27-mm) pine outside corner molding
Waterproof premium glue
Weatherproof 30-minute clear silicone
Wood putty
80-grit sandpaper
1-in. (25-mm) finish nails or galvanized wire nails
#18 (¾-in./20-mm) escutcheon pins
Round dowel, ⅜ in. (1 cm) in diameter
Round dowel 5⁄16 in. (8 mm) in diameter
Square ½ x ½ in. (12 x 12 mm) dowel
Coral, blue, and white exterior paint
Four #4 (2½-in./60-mm) zinc-plated steel screw eyes
10 ft (3 m) #16 (1.6 mm) single jack zinc-plated chain
Basic tool kit (see page 126)
Needle-nose pliers

finished size

Approx. 11½ x 12¾ x 9 in. (29 x 32 x 23 cm), excluding the chain

cutting list

Front and back: 10 x 5 ½ in. (25 x 14 cm)—cut 2 from dog-ear fence board
Feeder base: 11½ x 5½ in. (29 x 14 cm)—cut 1 from dog-ear fence board
Long sides of feeder: 11½ x 2 in. (29 x 5 cm)—cut 2 from furring strip board
Short sides of feeder: 6¾ x 2 in. (17 x 5 cm)—cut 2 from furring strip board
Roof: 9 x 5½ in. (23 x 14 cm)—cut 2 from dog-ear fence panel

4 Paint the exterior of all the panels with coral paint, leaving the insides raw. Remember to block up the entrance hole with wadded-up paper or tape, so that no paint can accidentally drift into the interior. Let dry.

5 Aligning the bottom edges and making sure they are centered, glue and nail the front and back panels to the long sides of the feeder trough.

6 Paint all sides of the roof panels and the reserved triangles from step 1 blue. Let dry.

7 Cut a 9-in. (23-cm) length of 3/8-in. (1-cm) round dowel. Insert it through the holes drilled in step 3 to create an internal perch.

8 Apply glue to the peak of the roof, then line up the roof pieces with the edges of the roof peak, making sure that there is the same amount of overhang front and back. There will be a V-shaped gap between the two roof pieces at the apex. Nail in place.

9 Arrange two blue triangles on the front of the house section. When you're happy with the position, glue them in place. Drill three 1/16-in. (1.5-mm) pilot holes (see page 131) in each triangle, then gently hammer in escutcheon pins. Repeat on the back of the house section.

10 Cut four 1½-in. (4-cm) lengths of round 5/16-in. (8-mm) dowel. Referring to the photograph on the right, drill two 5/16-in. (8-mm) holes for dowel perches in the front of the house section and two in the back. Gently hammer the dowels in place.

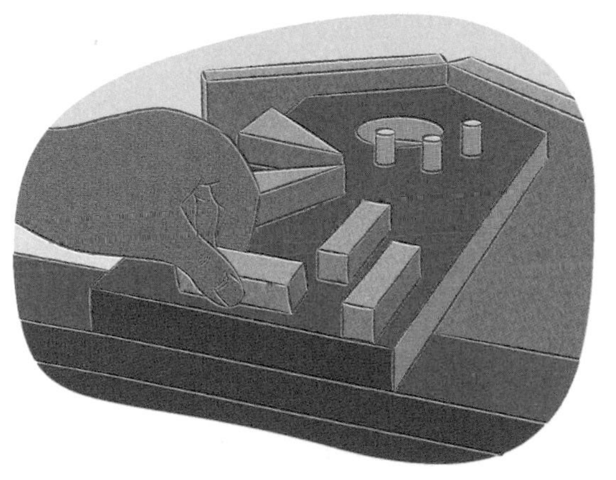

11 Cut six 2¾-in. (7-cm) lengths of square ½-in. (12-mm) dowel. Paint them blue on all sides and let dry. Arrange three on the front and three on the back, then glue and nail them in place, as you did in step 9 with the triangles, using two escutcheon pins for each one.

12 Paint the dowels on the outside of the feeder white. Let dry.

13 Cut a 9½-in. (24-cm) length of outside corner molding. Paint it coral and let dry. Silicone it into the V-shaped gap at the roof apex. Let dry. Cut a 9½-in. (24-cm) length of 3/8-in. (1-cm) round dowel. Paint it white and let dry. Silicone it into the V-shaped crevice of the corner molding and let dry.

14 Referring to the photo above and using a 1/16-in. (1.5-mm) bit, drill a pilot hole at each corner edge of the roof. Insert a screw eye into each one. Cut the chain into two 28-in. (70-cm) lengths. Using needle-nose pliers, attach one end of each chain to the screw eyes on opposite corners, following the instructions on page 133. Use a single chain link to join the two chains together in the center for hanging.

Birds will be delighted to get out of the wind, rain, snow, and heat in this adorable shelter. Hand painted with grass and yellow flowers, it's equipped with three perching wires where birds can snuggle up together and talk about the weather.

bird shelter

European Starling

materials

Two 6ft x 5½ in. x ⅝ in. (180 x 14 x 1.5 cm) cedar dog-ear fence board
16½-in. (42-cm) length of ¾ x ¾ in. (20 x 20 mm) square dowel
Waterproof premium glue
Wood putty
80-grit sandpaper
1-in. (25-mm) and 1¼-in. (30-mm) finish nails or galvanized wire nails
Twelve 1½-in. (40-mm) wood joiner nails
Four ½-in. (12-mm) screws
18-gauge (1 mm) rebar tie wire
Green exterior spray paint
Blue, dark green, light green, white, yellow, and black craft paint
1-in. (25-mm) pipe flange
1-in. (25-mm) nipple pipe
Basic tool kit (see page 126)
Wire cutters
Needle-nose pliers

finished size

Approx. 17¾ x 15¾ x 6¼ in. (45 x 40 x 16 cm)

cutting list

Sides: 16½ x 5½ in. (42 x 14 cm)—cut 2, with 1 short end cut at a 15° angle
Back: 14½ x 5½ in. (37 x 14 cm)—cut 4 (reserve 1 for floor)
Roof: 15⅝ x 5½ in. (39.5 x 14 cm)—cut 1

1 Dry fit the three back panels and the two side panels together. Make a pencil mark ½ in. (12 mm) from the bottom of the lowest back panel. This will stick out ½ in. (12 mm) below the side panels.

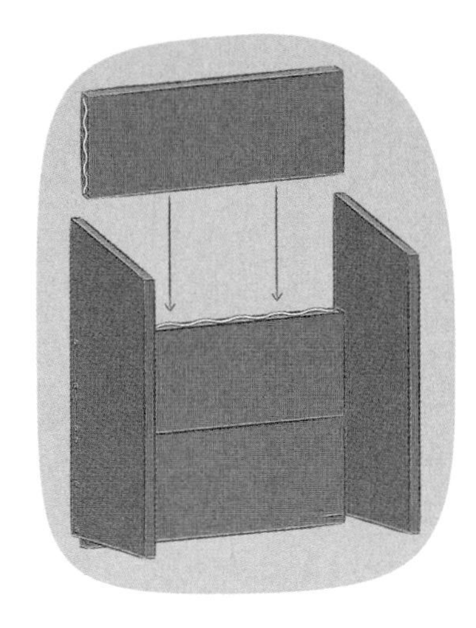

2 Apply glue to the edges of the lowest back panel, and nail on the side edges, aligning them with the marks on the back panel at the base. Attach the remaining two back panels to the sides in the same way, making sure they butt up against each other. When complete, the two side panels will extend ½ in. (12 mm) above the back panel.

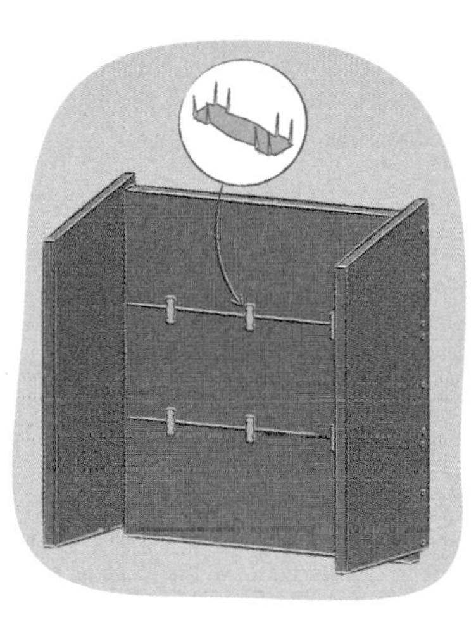

3 To ensure you have a good bond, hammer in 1½-in. (40-mm) wood joiner nails approx. 3½ in. (9 cm) apart on each "seam" of the back and on both sides.

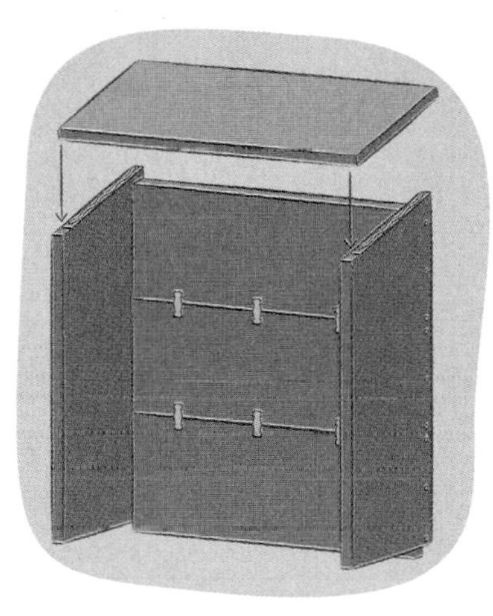

4 Measure and mark ¾ in. (2 cm) back from the front sloping edge. Glue and nail the roof panel in place, aligning it with your marks.

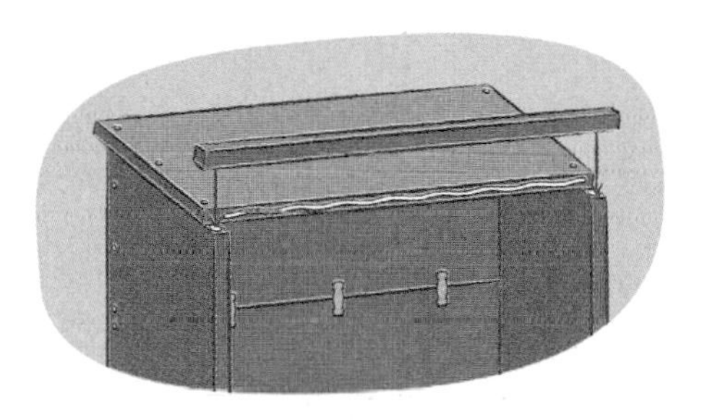

5 Cut a 16½-in. (42-cm) length of ¾-in. (20-mm) square dowel. Glue and nail it across the top front edge of the shelter, in line with your pencil marks, using 1¼-in. (30-mm) nails.

6 Along the bottom edge of the shelter, apply glue just above the overhang and on the side edges. Slide in the floor and nail it securely on the sides and back.

7 Spray paint the whole shelter green, inside and out, and let dry.

8 Now mark on both sides where the holes for the perching wires will go. Make one pair of marks 2½ in. (6 cm) from the back and 3 in. (7.5 cm) down from the top edge; the second pair 2½ in. (6 cm) from the back and 3 in. (7.5 cm) up from the bottom floor edge; and the third pair 2 in. (5 cm) from the front edge and 7 in. (19 cm) up from the bottom floor edge. Drill the holes using a 1/16-in. (1.5-mm) bit.

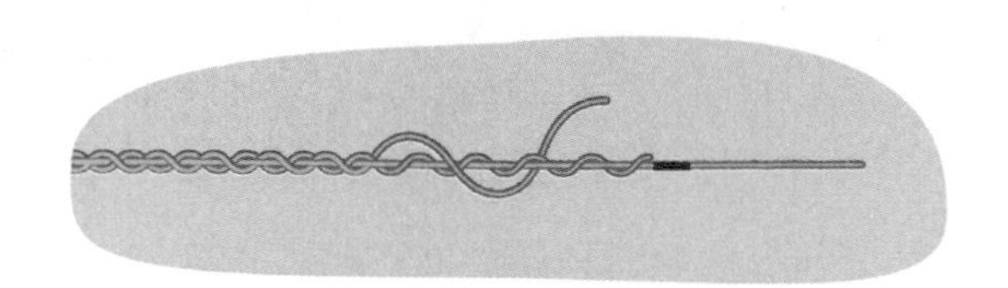

9 Cut one 19-in. (48-cm) and two 15-in. (38-cm) lengths of rebar wire. With red paint, make a mark 2 in. (5 cm) from each end of the longer wire. Working between the red marks, twist each of the shorter wires in turn around the longer one. Repeat twice more to get three thick perches, each with 2 in. (5 cm) of single wire at each end.

10 Using needle-nose pliers, pull the single wires through the holes drilled in step 8, pull taut, then twist the ends to secure.

11 Center the flange on the bottom of the shelter. Drill pilot holes (see page 131) for ½-in. (12-mm) screws and attach a flange plate for 1-in. (25-mm) nipple pipe. This will slide down an aluminum hollow fence post that can be planted in the ground with concrete.

12 Paint the shelter however you wish. I chose to make a garden design, with blue sky in the background. First, I drybrushed blue paint over the top two-thirds, inside and out. Then I dipped my brush into both light and dark green paint and applied single downward strokes to create long, stem-like greenery over the bottom third. I used the same technique with yellow and white paint, making small strokes in a circular pattern to create flowers. For the flower centers, I dipped the end of a nail in black paint and lightly pounced in a circular motion, making a cluster of tiny dots.

wine bottle feeder

What a clever way to recycle a glass bottle! Filled with birdseed and hand painted with dangling grapes, it is a lovely addition to the patio porch, where you can watch birds dine while sipping a glass of wine. Did I tell you how much I love birds (and wine)?!

materials

One wine bottle, cleaned and end cut off (see step1)
220-grit sanding sponge
One 6ft x 5½ in. x ⅝ in. (180 x 14 x 1.5 cm) cedar dog-ear fence board
Waterproof premium glue
Wood putty
80-grit sandpaper
1-in. (25-mm) finish nails or galvanized wire nails
1¼-in. (30-mm) exterior screws
Cranberry exterior spray paint
Green (three shades) and brown craft paint
Approx. 30 in. (75 cm) 18-gauge (1-mm) rebar tie wire
Miniature drawer pull (craft store)
Two antique corners (craft store)
One wine and one champagne cork
Plastic funnel
Basic tool kit (see page 126)
Tile saw
Wire cutters
Needle-nose pliers

finished size

Approx. 15 x 6½ x 7¾ in. (38 x 16.5 x 19.5 cm)

cutting list

Bottle support strip: 10 x 2⅜ in. (25 x 5.5cm) —cut 1
Back panel: 14 x 5½ in. (35.5 x 14 cm)—cut 1
Tray base: 7 x 5½ in. (18 x 14 cm)—cut 1
Tray sides: 7 x 1½ in. (18 x 4 cm)—cut 2
Tray front: 6⅝ x 1½ in. (17 x 4 cm)—cut 1

1 Wearing safety goggles and gloves, cut the base off the wine bottle, using a diamond blade on a tile saw. Wet sand the rough edges with a 220-grit sanding sponge. (Note that the sanding sponge as well as the bottle needs to be wet in order to smooth the sharp edges.)

2 Cut a 1¼-in. (32-mm) hole in a scrap piece of fence board that is roughly 5½ in. (14 cm) square. Then cut the fence board down to 2⅜ in. (5.5 cm) wide by 4½ in. (11.5 cm) long, centering the hole. This is the bottleneck holder.

Cutting a hole in a bigger piece of wood ensures it doesn't split apart. This will give enough room to make your hole, then cut the wood down to size.

Always drill pilot holes for the screws *(see page 131).*

3 The length of your cut bottle will determine where to place the bottleneck holder. Line up the cut end of the bottle with the top edge of the long support strip. Slide the bottleneck holder over the neck of the bottle and mark where it will be attached. Glue and nail the bottleneck holder to the support strip, then add a 1¼-in. (30-mm) screw in the center between the nails for extra support.

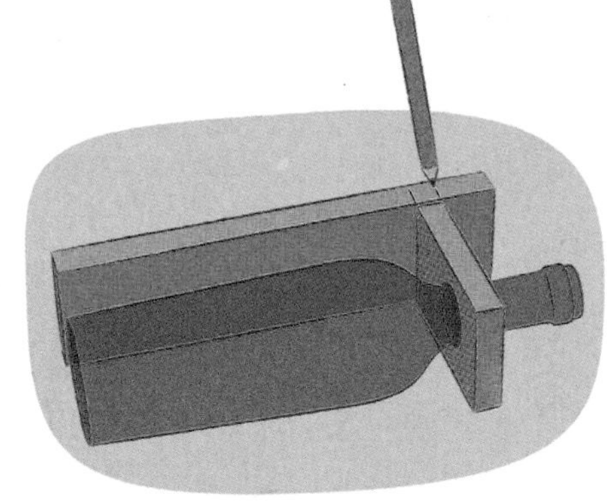

4 Place the bottle in the holder, then dry fit the back panel and bottom tray to see where the holder needs to sit in order for the seeds to flow correctly. The top of the bottle should be ¼ in. (6 mm) from the tray bottom and the support strip should be centered on the width of the back panel. Mark with pencil.

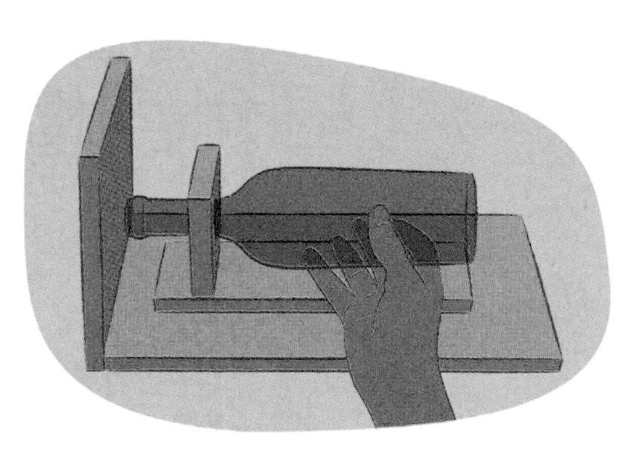

5 Glue and nail the support strip to the back panel, then add a couple of 1¼-in. (30-mm) screws from the back for extra support.

6 Place the tray base under the back panel. Glue and nail it in place, then add a 1¼-in. (30-mm) screw about 2 in. (5 cm) from each side.

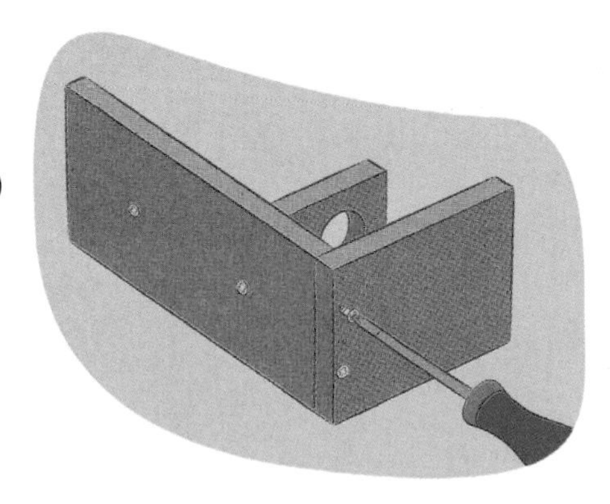

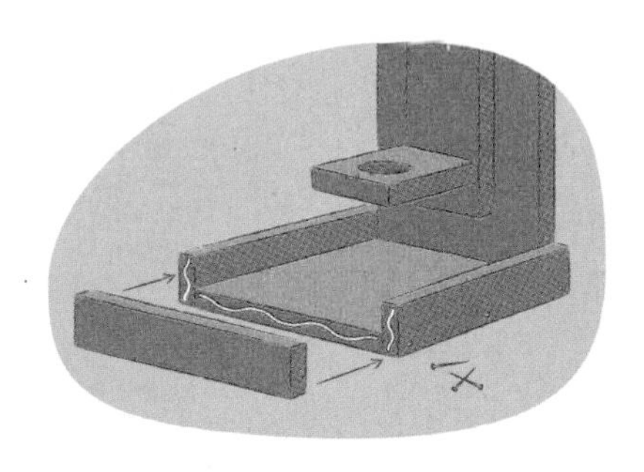

7 Glue and nail the tray sides to the base, then attach the tray front.

To cut the tray sides and front, cut a piece 7 in. (18 cm) long across the width of the fence board, then line up the short side with the saw fence and cut into three strips 1½ in. (4 cm) wide.

8 Paint everything except the interior of the feeder tray cranberry and let dry. Keep the feeder tray free of any paint by lining it with paper or a piece of wood.

9 On each side of the central bottle-holder panel, about ½ in. (12 mm) below the top of the panel, drill a 1⁄16-in. (1.5-mm) hole all the way through the back panel. Cut about 24 in. (60 cm) of wire. Form it into a loose spiral to prevent the bottle from tilting from side to side. Then feed the loose ends of the wire through the holes and twist them together on the back to form a loop for hanging.

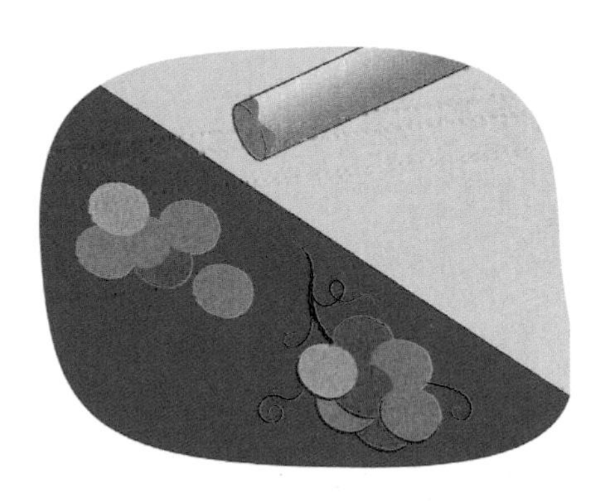

10 To paint the bunches of grapes, cut a ⅜-in. (1-cm) round dowel to pencil length, dip the end in green paint, and press it onto the wood. Use three different shades of green and layer the circles to create the 3D effect. Add dark green stems with a fine brush, then a few swirly lines here and there. Paint brown around some of the grapes for more detail.

11 The adornments are optional. I attached a miniature drawer pull and two antique corners to the top edge of the back. The cork decoration was made by drilling a ⅛-in. (3-mm) hole all the way through the cork, feeding through a length of wire, and twisting the ends into spirals. I used leftover wire to make an S-hook to attach the cork to the wires around the bottle.

12 Find a funnel that fits the cut end of your bottle perfectly, then paint it to match the feeder. Place a champagne cork inside. They act as a lid to keep out unwanted debris and prevent the birds from going in through the cut end of the bottle. Remove them when you want to fill up the feeder, then replace them.

Teacups make the cutest bird feeders! Who doesn't have an old set of teacups that Grandma passed down? Painted with tiny pink flowers and with a handcrafted shutter backsplash, not only is this bird feeder cute and functional, it will be the talk of the party.

Blue Tit

cup and saucer bird feeder

materials

Square ¾ x ¾-in. (20 x 20-mm) dowel
One 6ft x 5½ in. x ⅝ in. (180 x 14 x 1.5 cm) cedar dog-ear fence board
Waterproof premium glue
Wood putty
80-grit sandpaper
1-in. (25-mm) finish nails or galvanized wire nails
Ten 1¼-in. (30-mm) loose finish nails
Eight decorative 1¼-in. (30-mm) bronze-colored screws
Gray and white exterior spray paint
Teacup and saucer
Weatherproof 30-minute clear silicone
Old bronze-colored knob
Two steel D-ring picture hangers
Pink, green (two shades), and yellow craft paint
Floral stamp
Basic tool kit (see page 126)

finished size

Approx. 13¾ x 7 x 6 in. (35 x 18 x 15 cm)

cutting list

The shutter frame is cut from square dowel. All other pieces are cut from dog-ear fence board.

Shutter frame: Cut two 7-in. (18-cm) and two 4¼-in. (11-cm) lengths
Shutter slats: 4½ x 5½ in. (11.5 x 14 cm)—cut 1 (see step 3)
Shutter plate rest: 5¾ x 5½ in. (14.5 x 14 cm)—cut 1
Shutter back: 12½ x 5½ in. (32 x 14 cm)—cut 1 *
Feeder trough base: 5¾ x 5½ in. (14.5 x 14 cm)—cut 1 **
Feeder trough front: 7 x 1¾ in. (18 x 4.5 cm)—cut 1
Feeder trough sides: 5½ x 1¾ in. (14 x 4.5 cm)—cut 2
Cup "pedestal": 1¾ x 1¾ in. (4.5 x 4.5 cm)—cut 1
Roof: 5¾ x 3⅝ in. (14.5 x 9 cm)—cut 1

notes

* Cut the shutter back after you've assembled the shutter and plate rest so you can be sure it's exactly the right size (see step 6).
** You may need to adjust the size of the feeder trough slightly to fit your cup and saucer.

1 Apply silicone to the side of the cup opposite the handle. Then stick the cup to the saucer, placing it so that the design will show. Let dry for 45–60 minutes before handling. Set in a place where it won't get knocked over and broken.

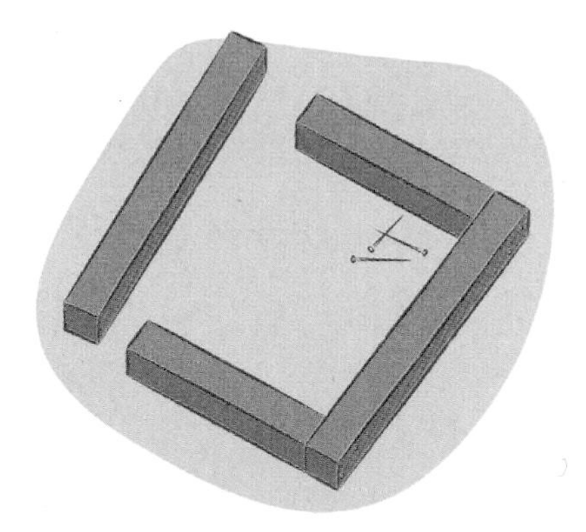

2 Glue and nail the short shutter-frame pieces to one long piece, nailing twice on each end so that they won't move. Leave the other long end open so that you can insert the slats.

If you feel your fingers are getting too close to the spinning blade, it's better to cut another piece of fence board and cut the remaining slats from that.

3 Take the piece of board that you cut for the shutter slats. Place the cut end against the saw fence and cut five pieces ¾ in. (2 cm) wide and 4½ in. (11.5 cm) long.

4 Line up the slats inside the frame, spacing them evenly, with the ¾-in. (2-cm) face against the frame and the ⅝-in. (1.5-cm) edge facing upward. Once you've lined them up, carefully mark the frame to show where the slats will sit so that you will know where to nail. Mark the unattached 7-in. (18-cm) side of the frame in the same way, so that the slats will line up perfectly. Using a 1⁄16-in. (1.5-mm) bit, drill a pilot hole (see page 131) in the center of each marked section.

5 Apply glue to the short sides of the frame, then nail the remaining long side in place. **Do not glue or nail the slats**—you need to be able to move them so that you can line them up with your marks. Once you're happy with the alignment, drive in 1¼-in. (30-mm) loose finish nails to attach the slats, angling them slightly.

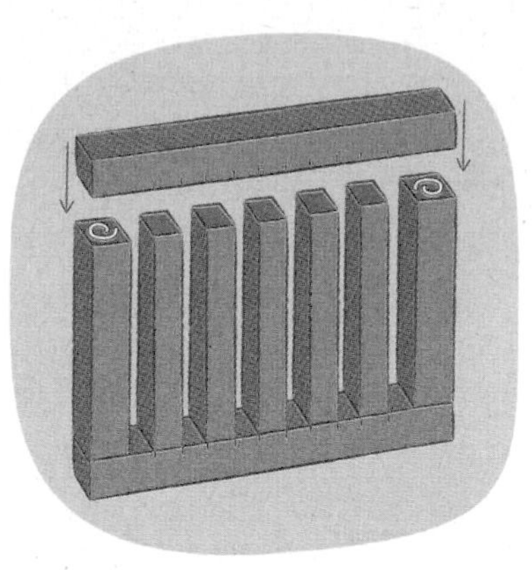

6 Place the plate rest against the shutter, on top of uncut fence board, and mark where to cut the back piece; it will be approx. 12½ in. (32 cm) long.

7 Spray the entire piece gray on both sides. Before the paint dries, spray white paint over the top to get the cool "shabby chic" look. Let dry. Paint the cup "pedestal" and the roof in the same way and set aside.

8 Apply glue to the top and bottom edges of the shutter and line it up with the top edge of the back panel. Using a ⅛-in. (3-mm) bit, drill pilot holes in the short sides of the shutter, 1⅛ in. (3 cm) from the long edges, then drive in 1¼-in. (30-mm) decorative screws. Attach the plate rest in the same way, making sure it is snug against the bottom of the shutter and positioning the screws ½ in. (12 mm) from the edges.

9 Glue and nail the feeder trough sides to the base, then glue and nail on the front. Paint the outside of the trough in the same way as the shutters.

10 Apply glue to the bottom of the shutter back and slide it in between the sides of the feeder trough. Nail from the bottom and on each side.

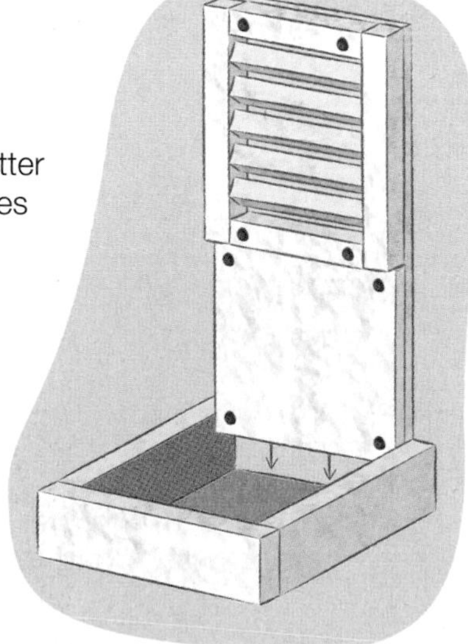

11 Glue and nail the cup pedestal inside the feeder trough 1½ in. (4 cm) from the front and 2 in. (5 cm) from each side. Apply silicone to the bottom of both the cup and saucer and carefully stick them to the pedestal and plate rest. Let the silicone dry for 45 minutes.

12 Apply glue to the top of the back and nail the roof in place.

You can find this kind of knob in any hardware store, just make sure you buy one that comes with a threaded screw, not a screw that is attached to the knob.

13 Drill a 3⁄16-in. (5-mm) hole through the center of the roof and insert the knob, securing with the screw.

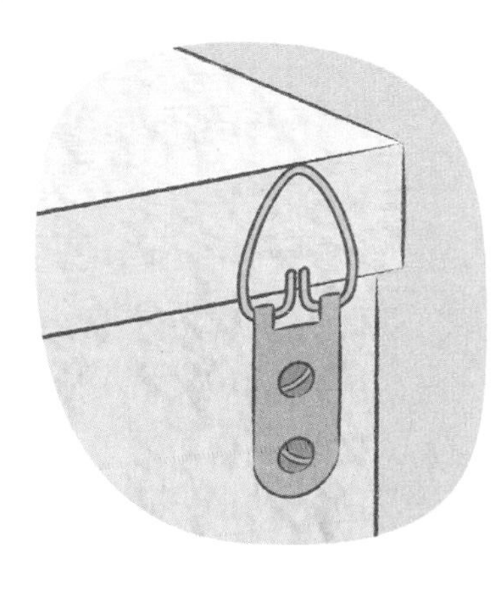

14 Line up the top point of the D-ring loops with the edge of the roof and mark the screw holes. Using a ⅛-in. (3-mm) bit, drill pilot holes and insert ½-in. (12-mm) screws.

15 Decorate the feeder with trailing vines of pink flowers (see page 139). I used a floral stamp, but you could paint them freehand if you prefer.

An octagonal basket that can either hold another basket (as shown) or be hung on the wall, filled to the brim with nest-building materials, this is a cool and easy-to-build project. Fill it with straw, Spanish moss, strips of paper, yarn, cotton fabric, twigs, dog hair, cat hair, grass, sticks… You never know—it may even serve as a nesting spot for non-cavity-nesting birds.

hanging material basket

materials

One 6ft x 5½ in. x ⅝ in. (180 x 14 x 1.5 cm) cedar dog-ear fence board
One garden stake, 36 x 2½ x ¾ in. (100 x 6 x 2 cm)
Waterproof premium glue
Wood putty
80-grit sandpaper
1-in. (25-mm) finish nails or galvanized wire nails
Four 1¼-in. (30-mm) exterior screws
One 1-in. (25-mm) bronze screw
Six decorative 1¼-in. (30-mm) screws
Six ½-in. (12-mm) screws
Red exterior spray paint
Wire basket, 9 in. (23 cm) in diameter and 3 in. (7.5 cm) high
Three old rusty springs
18-gauge (1-mm) rebar tie wire
Nesting materials (see step 9)
Basic tool kit (see page 126)
Wire cutters
Needle-nose pliers

finished size

16 x 16 x 5½ in. (40 x 40 x 14 cm)

cutting list

Basket sides: 6 x 5½ in. (15 x 14 cm)—cut 8, beveling each long side of each piece at 22.5˚, with the bevels going in the same direction *

note

* These pieces will be 6½ in. (16.5 cm) long from end to end and 6 in. (15 cm) long on the inside bevel cut.

1 Glue and nail one beveled end to another until you have a complete octagon. It's simpler to nail at an angle on the top edge, to connect the pieces; once they're all joined together, nail on each long side panel.

Remember to hold the nail gun at an angle, so that the nails shoot into the wood and don't protrude on either side.

2 Paint it in the color of your choice, inside and out, and let dry.

3 Cut a 14½-in. (37-cm) length of garden stake for the inside of the box. This will give the box more strength and you can attach a hook to it if you want to hang the basket on a wall. Place it across the box, at the seam between the first and second pieces from the top, flush with the edge. Using a ⅛-in. (3-mm) bit, drill pilot holes (see page 131) and drive in 1¼-in. (30-mm) screws. Paint red and let dry.

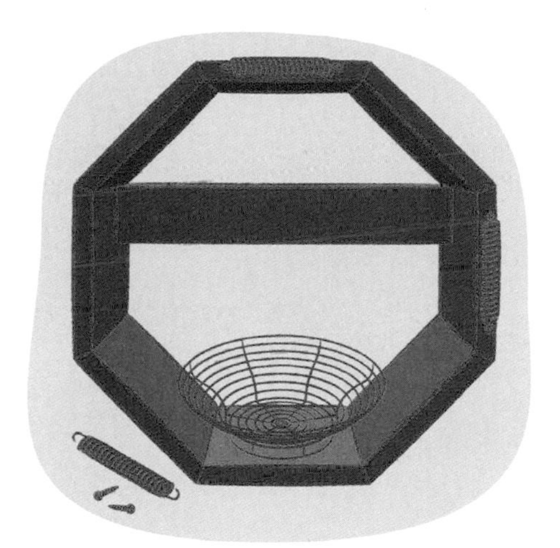

4 Add your basket, if using. Mine already had a hole in the base that I used to screw it in place with a 1-in. (25-mm) screw. Add bedsprings or other decoration to the side; I found these at an antiques store and attached them with decorative 1¼-in. (30-mm) bronze screws as close as possible to the inner side of the spring loop to keep them as tight and secure as possible.

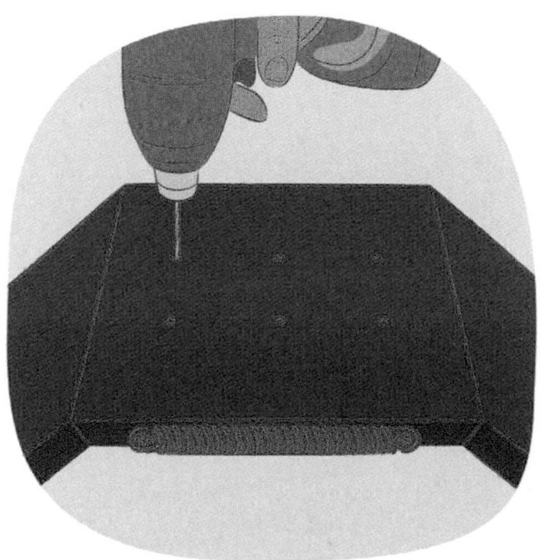

5 Using a pin bit, make three rows of two holes each in the top panel, with the outer holes 2 in. (5 cm) from the long edges of the panel and 1½ in. (4 cm) from the beveled seam edge and the remaining two holes centered between them.

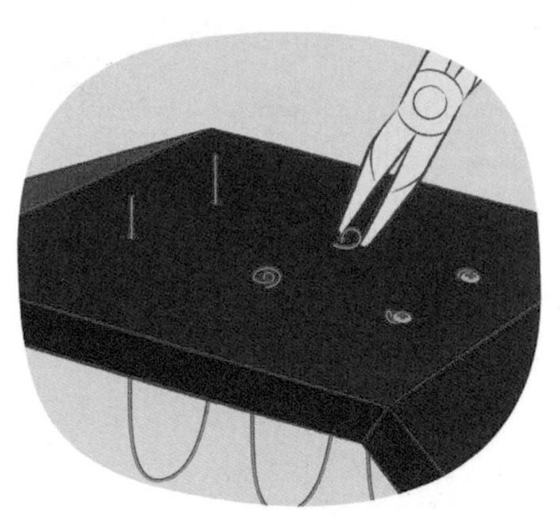

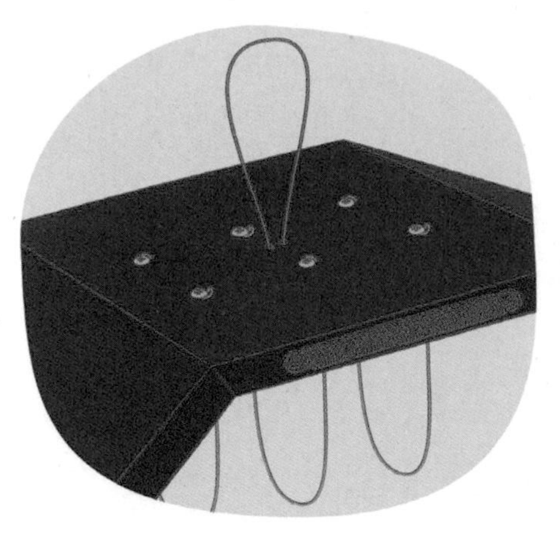

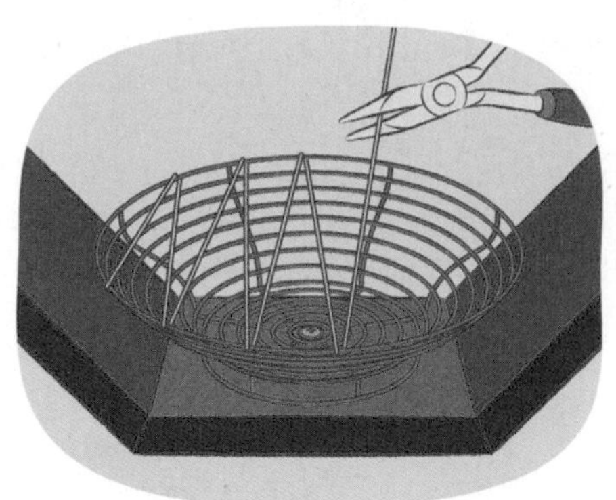

6 Cut three 16-in. (40-cm) lengths of 18-gauge (1-mm) rebar tie wire. Bend each one into a U-shape and push the ends through a pair of holes from the underside, leaving about 2 in. (5 cm) sticking out. Twist the ends into small coils to hold in place, then insert ½-in. (12-mm) decorative screws to secure the twisted ends.

7 Cut a 10-in. (25-cm) length of rebar tie wire. Using a pin bit, drill two holes ¼ in. (6 mm) apart in the center of the top panel, between the rows of holes. Insert the wire through the holes from the top, then twist the ends together on the inside of the octagon to form a hanging loop.

8 String 18-gauge (1-mm) rebar tie wire through your basket in order to keep the different nesting materials in place, twisting the ends around the basket lip.

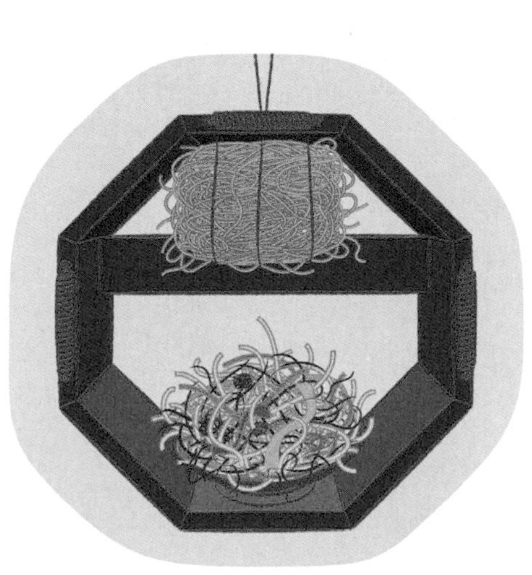

9 Cut a skein of yarn in half, then feed some of it through the three loops, teasing it out so that it's not all in one clump. Fill the basket with strands of raffia, moss, strips of cotton fabric, cat or dog hair, grass and anything else that might take the birds' fancy as nest-building material.

bird perch swing

I'm lucky enough to have hummingbirds visit my garden in California and I just love watching them, but they do need a place to land and rest between their 15-minute feeding times. I designed this pretty swing to give them somewhere to perch near the feeding bottle—but it would work just as well for other small birds.

Hummingbird

materials

Decorative wood appliqués, available from craft stores or online
One 6 ft x 3 in. x 1 in. (180 x 7.5 x 1.5 cm) common board (whitewood) or scrap from cedar fence board cut to size
48-in. (120-cm) length of ¾ x ¾-in. (2 x 2-cm) square dowel
Waterproof premium glue
Wood putty
80-grit sandpaper
½-in. and 1¼-in. (12-mm and 30-mm) finish nails or galvanized wire nails
Blue exterior spray paint
White, pink, and green craft paint
18-gauge (1-mm) heavy-duty garden wire
Miniature drawer pull, available from craft stores
Basic tool kit (see page 126)
Wire cutters
Needle-nose pliers

finished size

Approx. 10¾ x 11 x ¾ in. (27 x 28 x 2 cm)

1 Lay out your decorative wood embellishments to work out how big the panel needs to be, then cut it from leftover fence board; mine was 9⅝ x 3 in. (24.5 x 7.5 cm). Apply glue to the back of the wood embellishments, drill pilot holes with a pin bit (see page 131), and hammer in ½-in. (12-mm) nails to secure.

If your wood embellishments take up more or less space than mine, adjust the size of the fence board and dowel pieces accordingly.

2 Cut two 9⅝-in. (24.5-cm) lengths of square dowel. Glue and nail them to the top and bottom of the decorated panel, using 1¼-in. (30-mm) nails.

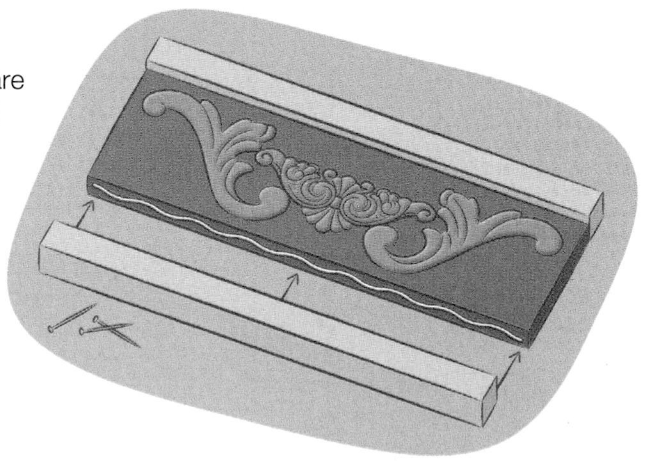

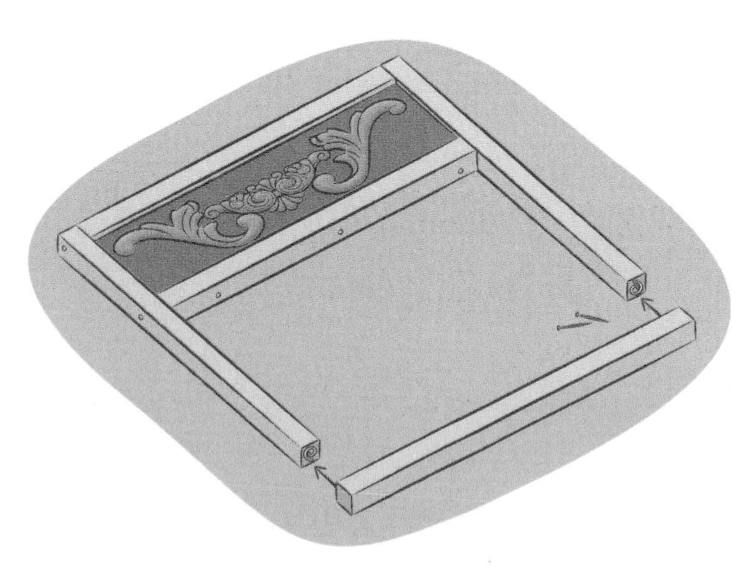

3 Cut two 10-in. (25-cm) lengths of square dowel and glue and nail them to the sides of the panel. Then cut an 11-in. (28-cm) length of square dowel and glue and nail it across the bottom to complete the "frame."

4 Paint the entire piece blue and let dry, then drybrush white paint over the top. Referring to the photos above, add trailing vines of green leaves and little pink flowers (see page 139).

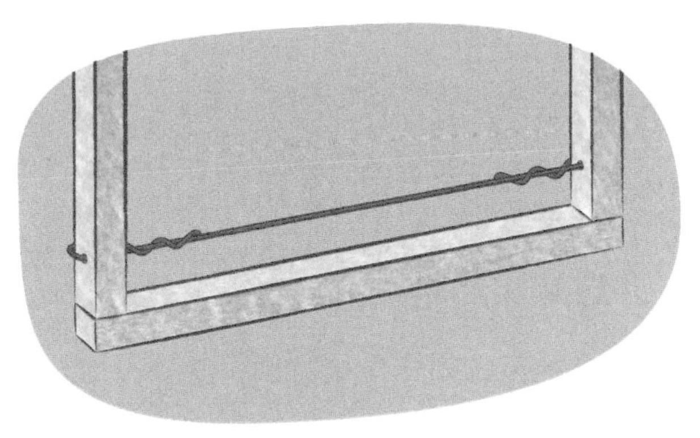

5 Near the base of the frame, drill a $\frac{3}{16}$-in. (5-mm) hole about 2 in. (5 cm) up on each side, making sure they are level. Cut a 16-in. (40-cm) length of heavy-duty wire. Feed it through the holes, pull it taut, and then bring the ends around the back of the frame and twist them around the stretched wire.

6 Drill pilot holes for the miniature drawer pull on top of the swing, then screw in the screws with a small screwdriver.

chapter 3

bees, butterflies, and ladybugs

Bees, butterflies, and ladybugs are such an important part of the ecosystem that I've created homes for these buzzing, fluttering insects to take refuge in. Plant these adorable homes in your garden, along with flowers, fruit trees, and vegetables, and watch how these simple creations make a difference.

Mason Bee

Keeping native non-stinging mason bees is a surprisingly easy way to help the environment, and it's also an inexpensive and educational project for kids. The best news is that these hyper-efficient pollinators will do wonders for your fruit crops and gardens!

cozy bee hotel

materials

One 6 ft x 5½ in. x ⅝ in. (180 x 14 x 1.5 cm) cedar dog-ear fence board
Waterproof premium glue
Wood putty
80-grit sandpaper
1-in. (25-mm) finish nails or galvanized wire nails
Weatherproof 30-minute clear silicone
Mason bee replacement nesting tubes (available online)
Wine corks
Moss or crinkle-cut shredded paper
Decorative exterior screws (optional)
16-in. (40-cm) length of 18-gauge (1-mm) wire (optional)
Basic tool kit (see page 126)

optional items for decoration

Green and red craft paint
3 metal bee adornments
#18 (¾-in./20-mm) gold escutcheon pins
Driftwood

finished size

Approx. 10¾ x 10¼ x 7½ in. (27.5 x 26 x 19 cm)

cutting list

Roof panel 1: 6 x 5½ in. (15 x 14 cm)—cut 1
Roof panel 2: 6½ x 5½ in. (16.5 x 14 cm)—cut 1
Roof supports: 3½ x 2½ x 2½ in. (9 x 6 x 6 cm) triangles—cut 2 (make a mark 2½ in./6 cm from the corner of the top and side edges of the fence panel and then join the marks to make a right-angle triangle)
Side walls: 6 x 5½ in. (15 x 14 cm)—cut 2, with the short edges cut at a 22.5° bevel and the bevel edges on each end going in same direction
Bottom platform: 6 x 5½ in. (15 x 14 cm)—cut 1
T-bar divider 1: 3½ x 5½ in. (9 x 14 cm)—cut 1, with both short ends cut at a 45° bevel and the bevels going in opposite directions
T-bar divider 2: 6¼ x 5½ in. (15.8 x 14 cm)—cut 1, with both short ends cut straight
Roof eave 1: 6⅝ x 5½ in. (16.8 x 14 cm)
Roof eave 2: 7¼ x 5½ in. (18.4 x 14 cm)

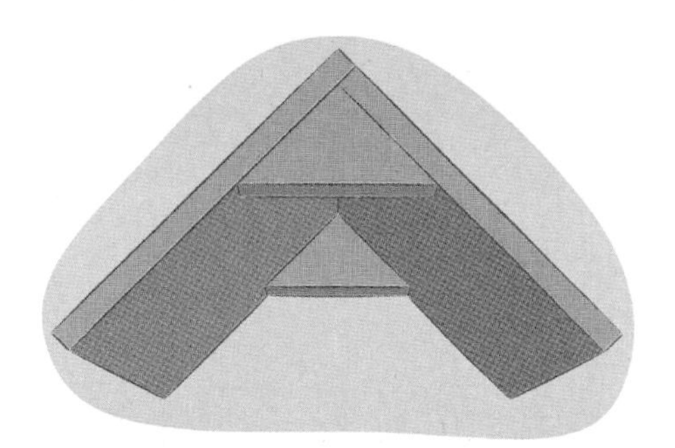

1 To make the roof, glue and nail roof panel 2 on top of roof panel 1 along one 5½-in. (14-cm) edge. When assembled, each side of the roof will be the same size. Glue and nail a roof support triangle at each inner corner.

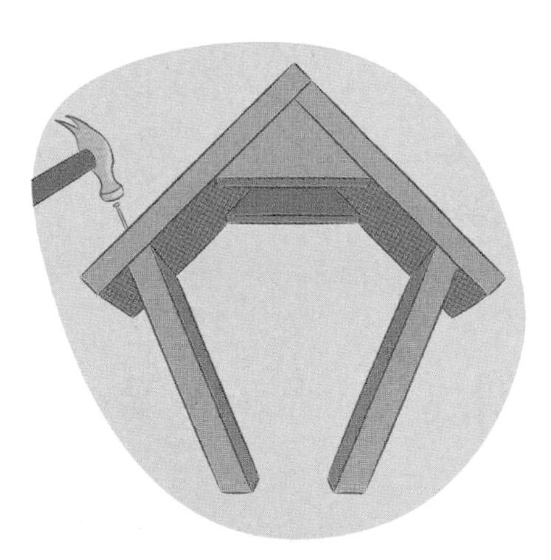

2 Glue and nail the side walls 1½ in. (4 cm) in from the edge of the roof, nailing them at an angle that follows the alignment of the wall piece so that the nail tip does not protrude and split the wood.

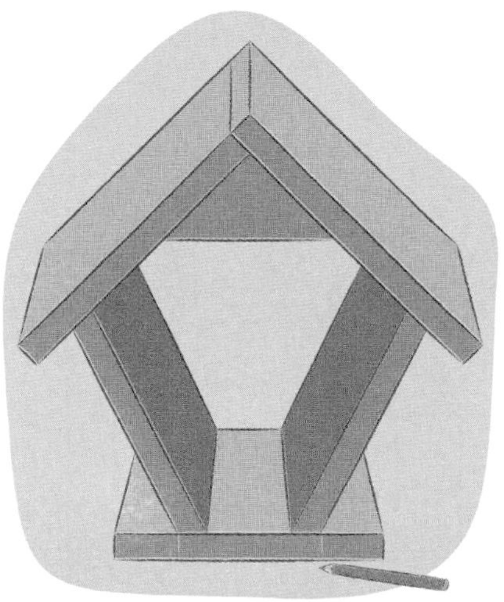

3 Glue and nail the bottom platform in place, aligning it with the front and back edges and making sure that the same amount protrudes at each side. (Mark each end with a pencil to ensure it lines up properly.)

4 Center T-bar divider 1 on top of T-bar divider 2, glue, and nail in place. Then install the divider in the bee house, gluing and nailing it in at the bottom and underside of the top, front and back, and nailing the top piece at an angle.

5 Place eave panel 1 on the roof 2 in. (5 cm) from the back, aligning it with the top and bottom of the roof. Glue and nail in place. Repeat with eave panel 2 on the opposite side, again aligning it with the top and bottom of the roof.

6 If you wish, paint the roof and base one color (I used red) and the body another (I used green). Alternatively, you could leave it unpainted to age naturally.

7 Fill the "hotel" with mason bee nesting tubes (available online), interspersing wine corks randomly, and stick them in place with silicone; you will need to cut ¼ in. (6 mm) off the tubes for them to fit flush in the chamber. Let dry for 45 minutes, then fill in the gaps with moss or crinkle-cut shredded paper, again securing it with silicone.

8 If you wish, add extra decorations such as metal bees (stick them in place with silicone, wipe off any excess with a cotton swab, then secure with escutcheon pins) or pieces of driftwood. Add some pieces of moss to the driftwood.

9 Place the bee hotel on top of a 4 x 4 post in the center of your flower garden (see page 135), or attach a wire to the back to hang it against a fence (see page 132).

This is a fun and easy project that small bees will love—a block created from raw wood with holes drilled to accommodate their nesting of eggs and larvae. Decorated with natural driftwood, it is sure to attract our friendly pollinators.

bee house

materials

One 6 ft x 5½ in. x ⅝ in. (180 x 14 x 1.5 cm) cedar dog-ear fence board
Waterproof premium glue
Wood putty
80-grit sandpaper
1-in. (25-mm) and 1½-in. (40-mm) finish nails or galvanized wire nails
Green exterior paint
Six or seven driftwood pieces, 3–6 in. (7.5–15 cm) in length
Weatherproof 30-minute clear silicone
Moss
Saw-tooth picture-frame hook (optional)
Basic tool kit (see page 126)

optional items for decoration

Three metal bee adornments
Three #18 (¾ -in./20-mm) escutcheon pins

finished size

9½ x 7½ x 4¾ in. (24 x 19 x 12 cm)

cutting list

Four-layer block: 8 x 5½ in. (20 x 14 cm)—cut 4
Roof piece 1: 3½ x 4¾ in. (9 x 12 cm)—cut 1
Roof piece 2: 3½ x 5½ in. (9 x 14 cm)—cut 1
Base: 3¾ x 5½ in. (9.5 x 14 cm)—cut 1

1 Cut one short end of each block piece at a 45° angle on both sides to form a peak. Glue and nail two blocks together, nailing from the back and nailing around the edges only (not in the center), so that you don't hit nails when you drill the holes in the next step. Continue until all four pieces have been joined together.

2 Using a 5⁄16-in. (8-mm) drill bit, drill approx. 50 holes randomly into the block, making them 1½ in. (4 cm) deep and spacing them roughly ¼ in. (6 mm) apart.

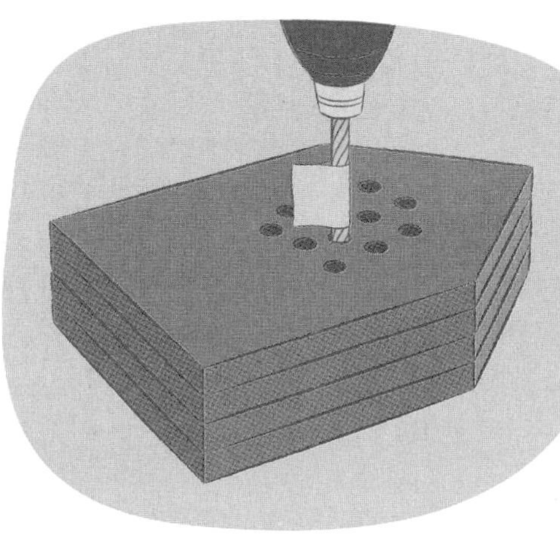

Stick a piece of masking tape 1½ in. (4 cm) from the drill bit tip so that you know when you've drilled to the correct depth.

3 To make the roof, glue and nail roof panel 2 on top of roof panel 1 along one 3½-in. (9-cm) edge. When assembled, each side of the roof will be the same size. Then glue and nail the roof to the block top by lining up the back edge of the roof with the back edge of the block; the roof will overhang at the front.

4 Glue and nail the base to the bottom of the block, aligning it with the back edge; the front will stick out by 1 in. (2.5 cm).

5 Paint the roof and base in your chosen color.

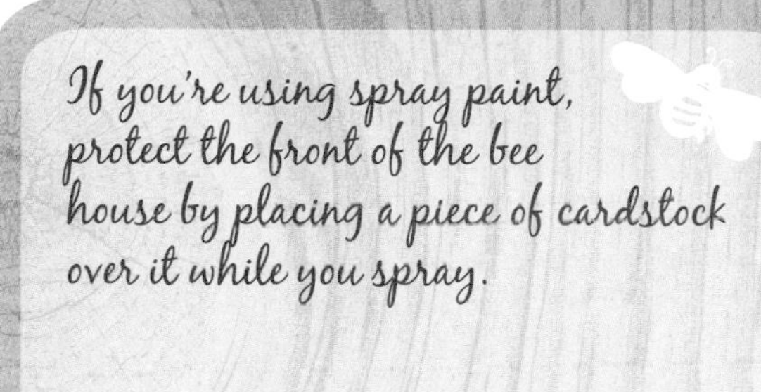

If you're using spray paint, protect the front of the bee house by placing a piece of cardstock over it while you spray.

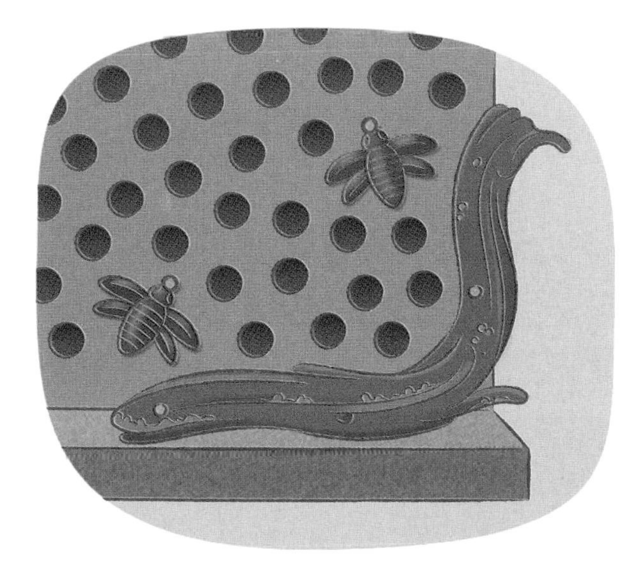

6 Attach the driftwood sticks with silicone, then drill pilot holes in the sticks with a 1/16-in. (1.5-mm) bit and gently hammer in finish nails to secure them. Using silicone, add a little moss for an extra finishing touch. If you wish, add a few metal bee embellishments with a dab of silicone and escutcheon pins.

7 Place the bee house on a post or rock, or attach a saw-tooth hook to the back to hang it on a fence near flowers.

A simple square box with driftwood "logs," this is an easy project for you and the kids to be at one with nature and will ensure that future generations of bees continue to thrive.

mason bee log home

Mason Bee

materials

One 6 ft x 5½ in. x ⅝ in. (180 x 14 x 1.5 cm) cedar dog-ear fence board
Wooden stake, 18 x 3 x 1 in. (45 x 5 x 2.5 cm)
Waterproof premium glue
Wood putty
80-grit sandpaper
1-in. (25-mm) finish nails or galvanized wire nails
Green exterior spray paint (optional)
Box of multi-colored jumbo smoothie straws
Green moss
Driftwood sticks, 2–6 in. (5–15 cm) long
Weatherproof 30-minute clear silicone
Two metal bee adornments
Two gold-colored #18 (¾-in./20-mm) escutcheon pins
Basic tool kit (see page 126)

finished size

4 x 5 x 6 in. (10 x 12.5 x 15 cm)

cutting list

Base and top: 4½ x 5½ in. (11.5 x 14 cm)—cut 2
Sides: 2 x 5½ in. (5 x 14 cm)—cut 2
Back: 2 x 3¼ in. (5 x 8.5 cm)—cut 1

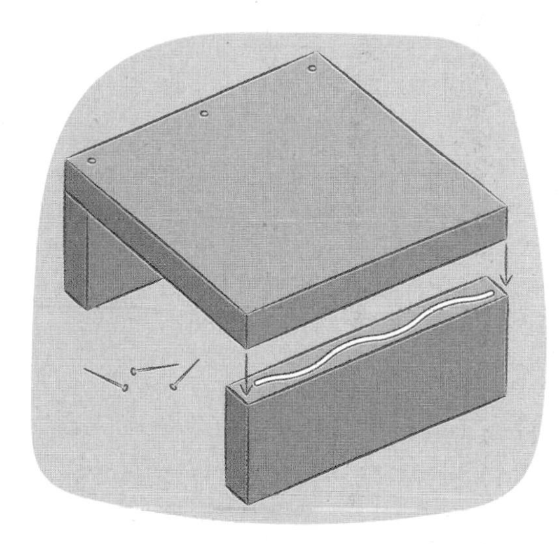

1 Glue and nail the sides to the base, leaving the top to one side for later.

2 Apply a line of glue along the back edge of the base, then slide in the back and nail it in place.

3 Paint the inside and outside of the box, including the top. Let dry. (This is optional—you can leave the box unpainted if you prefer.)

4 Cut the straws to 4¾ in. (12 cm) in length and place them in the box flush with the top. Glue and nail the box top in place. Fill any space left inside with off-cuts of the straws until the straws stay firmly in place.

5 Apply silicone over the top of the box and cover with green moss. Stick a few driftwood sticks in place with a line of silicone. Drill pilot holes (see page 131) in the sticks with a 1⁄16-in. (1.5-mm) bit, then gently hammer in finish nails to attach them to the top of the box.

6 Find the center of the back of the box, then glue and nail a stake to it.

7 Add driftwood sticks to the sides and front of the box, as in step 5, then fill in with moss as desired. Finally, add a couple of metal bee embellishments to the front, securing them with silicone and escutcheon pins.

butterfly feeding station

Butterflies are glorious creatures, graceful and spiritual. Reward these hard-working pollinators with their very own feeding station, where they can sip from nectar-soaked sponges to their heart's content. The custard cups also provide a place to add fruits such as slices of watermelon and orange.

materials

- One 6ft x 5½ in. x ⅝ in. (180 x 14 x 1.5 cm) cedar dog-ear fence board
- Waterproof premium glue
- Wood putty
- 80-grit sandpaper
- 1-in. (25-mm) finish nails or galvanized wire nails
- One 1¼-in. (30-mm) exterior screw
- Two ½-in. (12-mm) bronze screws
- Two 6-oz (170-g) glass custard cups, 4 in. (10 cm) in diameter
- Copper, blue, and yellow craft paint
- Scrap of square wood to paint squares
- Sea sponge, 3½ x 2½ in. (9 x 6 cm) or larger
- Ashland® natural raffia
- 16-in. (40-cm) length of 18-gauge (1-mm) wire
- Basic tool kit (see page 126)
- Compass
- Jigsaw

optional items for decoration

- Door plate, approx. 7½ x 2½ in. (19 x 6 cm)
- Butterfly metal knob, 2 in. (5 cm) wide
- Old-fashioned drawer pull
- Skeleton key adornment
- Metal keyhole adornment
- Gold or bronze ½-in. (12-mm) screws to attach your chosen decorations

finished size

13 x 8½ x 13 in. (33 x 21.5 x 33 cm)

cutting list

- **Center panel:** 11½ x 5½ in. (29 x 14 cm)—cut 1
- **Feeder tray base:** 12 x 5½ in. (30 x 14 cm)—cut 1
- **Feeder tray long sides:** 12 x 1½ in. (30 x 4 cm)—cut 2
- **Feeder tray short sides:** 6⅝ x 1½ in. (15.8 x 4 cm)—cut 2
- **Roof:** 6 x 5½ in. (15 x 14 cm)—cut 1, then cut in half lengthwise

Peacock Butterfly

My custard cups are 4 in. (10 cm) in diameter, with slightly sloping sides, so I made the radius of the circles 1½ in. (4 cm); you may need to adjust this if your custard cups are a different size.

1 Cut the center panel at a 45° angle on both sides to create a peak. Glue and nail a cut-off triangle to the center bottom of each side. This will help stabilize the center panel. Set aside.

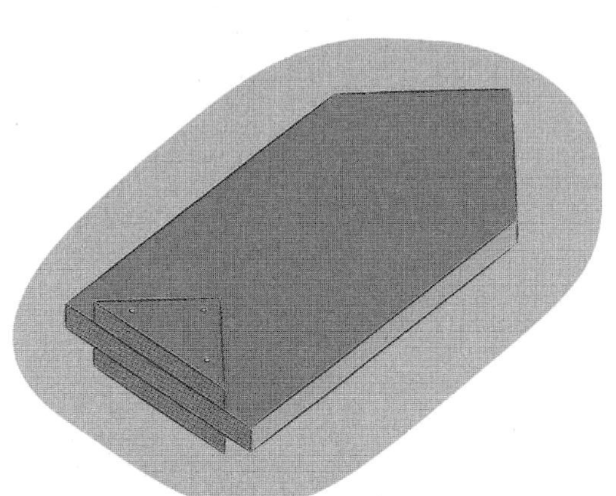

2 Draw a pencil line across the center of the feeder tray base to divide it into two squares, then mark the center of each square. Open out your compass to the required radius (see tip, above), then place the compass point on the center mark and draw the circles. Drill 5⁄16-in. (8-mm) holes all the way through at four points on each circle, then place your jigsaw blade in a hole to make the cut.

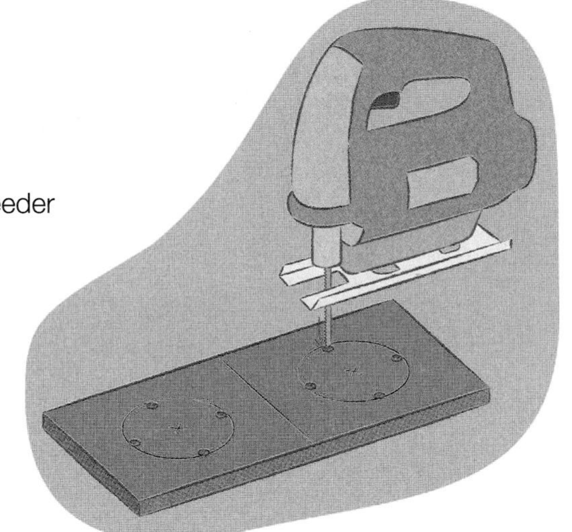

"nectar" recipe

1 part pure cane sugar to 5 parts water—it's as simple as that!

3 Mark the center of each long side of the feeder tray base, center the peaked panel on it, and glue and nail it in place. Drill a pilot hole (see page 131) in the center of the panel, then drive in a 1¼-in. (30-mm) screw.

4 Glue and nail the long side strips to the feeder tray base, then glue and nail the short side strips in place.

5 Cut ⅝ in. (1.5 cm) off the short side of one of the roof panels. Place the shorter roof panel level with the peak, with the same amount protruding on each side, and glue and nail it in place. Then glue and nail the longer roof panel to the opposite side of the peak.

6 Paint the feeder in your chosen color(s). I used copper for the roof and blue for everything else. Decorate the feeder as you wish: I used a door plate and a butterfly knob on the front, and a drawer pull, skeleton key, and keyhole adornment on the back.

7 Dip the end of a square piece of wood into yellow craft paint and press it over the sides of the feeder to form a random pattern.

8 Drill a pilot hole for ½-in. (12-mm) screws at each roof apex, then attach wire to form a hanging loop (see page 132).

9 Cut the sponge to fit the custard cups. Place the custard cups in the holes, with a sponge in each one, and tuck raffia in all around. When you're ready to place the feeder outside, soak the sponges in your homemade "nectar" solution (see opposite).

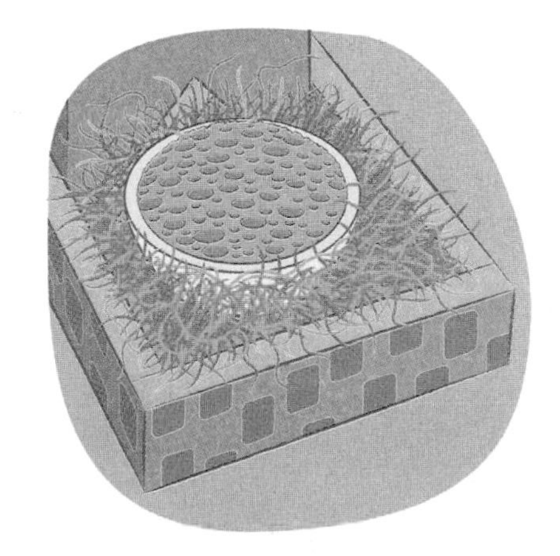

Painted white and filled with natural bark, driftwood, and moss, this butterfly house will look fabulous amongst the flowers and will give those winged angels a place to rest. Place it in the center of your flower bed, high above where butterflies are lingering, on a pole or post about 4 feet (120 cm) tall and planted 18 in. (45 cm) into the ground. Keep it away from bird feeders and birdhouses—we don't want the butterflies to be lunch!

Peacock butterfly

butterfly house

materials

Two 6ft x 5½ in. x ⅝ in. (180 x 14 x 1.5 cm) cedar dog-ear fence boards
One 6ft x 7½ in. x ¾ in. (180 x 19 x 2 cm) cedar dog-ear fence board
Wood putty
Waterproof premium glue
80-grit sandpaper
1-in. and 1¼-in. (25- and 30-mm) finish nails or galvanized wire nails
Four 1¼-in. (30-mm) exterior screws
Two 2¼ x 1-in. (5.7 x 2.5-cm) gold or silver hinges
One small latch
Gray and white exterior spray paint
Two or three pieces of bark, approx. 8 x 7 in. (20 x 18 cm)
Weatherproof 30-minute clear silicone
Pieces of driftwood 7–10 in. (18–25 cm) high
Moss
Basic tool kit (see page 126)
Jigsaw

finished size

Approx. 26½ x 6¾ x 5½ in. (67.5 x 17 x 14 cm)

cutting list

Front: 24 x 5½ in. (60 x 14 cm)—cut 1, beveling one short end at 22.5°
Sides: 26½ x 5½ in. (67 x 14 cm)—cut 2, mitering one short end of each one at 22.5°, so that one side measures 26½ in. (67 cm) and the other 24 in. (60 cm)
Upper back: 5¾ x 5½ in. (14.5 x 14 cm)—cut 1, beveling one short end at 22.5°
Floor: 4¼ x 5½ in. (11 x 14 cm)—cut 1
Lower back: 20⅛ x 5½ in. (51 x14 cm)—cut 1
Roof: 7½ x 7½ in. (19 x 19 cm)—cut 1 *
Bark stopper: 3 x 5½ in. (7.5 x 14 cm)—cut 1

note

* Cut the roof from fence board that is 7½ in. (19 cm) wide; all other pieces are cut from 5½-in. (14-cm) wide board.

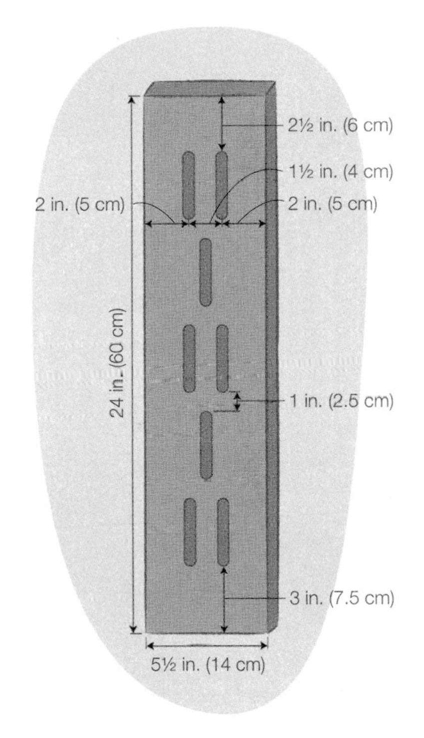

1 Mark on the front with a pencil where the entrance slots are going to be. Starting 2½ in. (6 cm) down from the top beveled edge, mark five rows of 3 x ½-in. (7.5 cm x 12-mm) vertical slots, as shown in the diagram. There should be 1 in. (2.5 cm) between each row; where there are two slots in a row, they should be 1½ in. (4 cm) apart. You should end up with 3 in. (7.5 cm) on the bottom.

2 Using a ½-in. (12-mm) hole saw, drill out holes at each end of your marks. Using a jigsaw, saw out the slots between the holes, then sand the splinters away with 80-grit sandpaper.

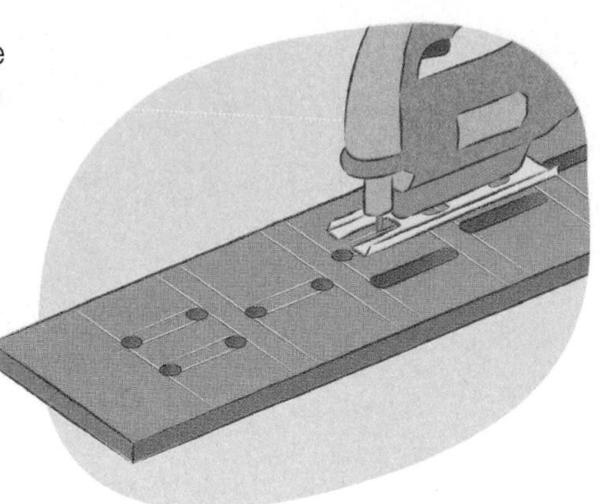

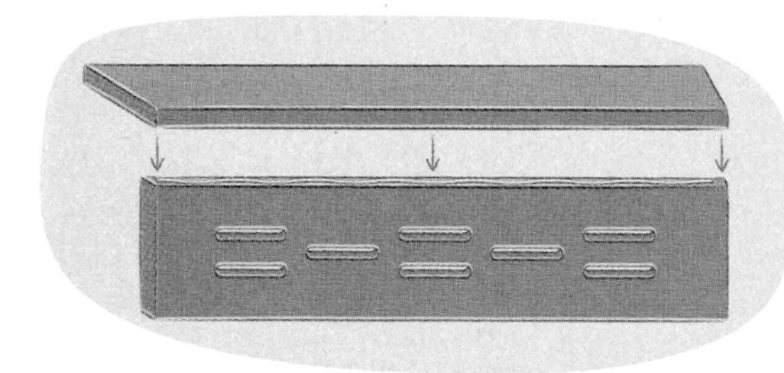

3 Glue and nail the 24-in. (60-cm) edge of the side panel on top of the front panel, aligning the panels top and bottom. Repeat on the opposite side.

4 Apply glue to the top 5¾ in. (14.5 cm) of the side panels, then slide the upper back panel in and nail it in place, aligning the top of the beveled edge with the highest point of the side panels. This will provide a place to attach the hinges for the swing-out door.

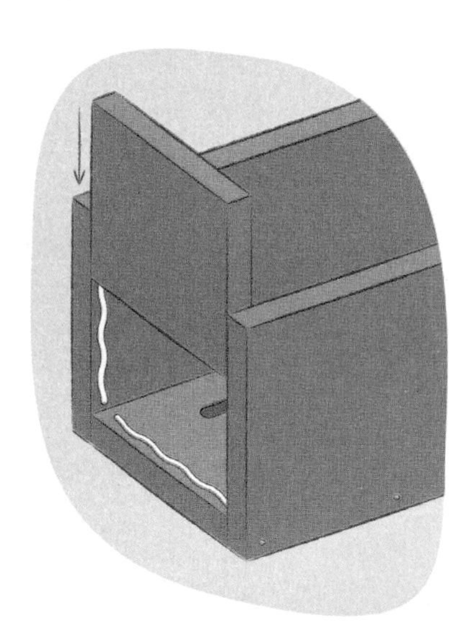

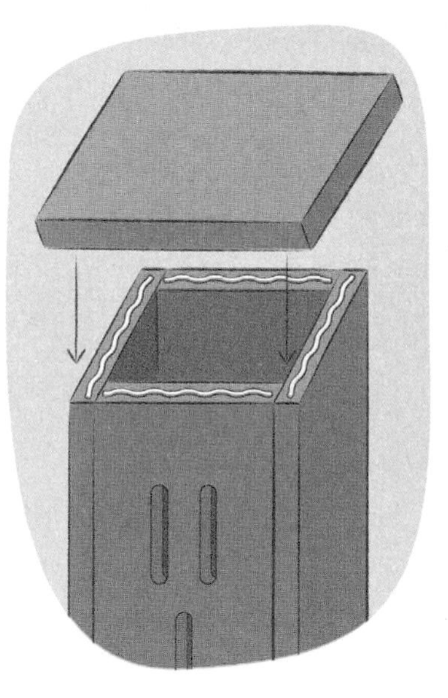

5 Dry fit the floor to ensure that it will sit flush with the edges and adjust if necessary. Glue and nail the floor in place.

6 Dry fit the lower back panel to ensure it sits flush to the floor. Drill pilot holes (see page 131) for the hinges in the back panels, with one half of each hinge on the upper back panel and the other half on the lower back panel, then drive in 1¼-in. (30-mm) screws.

7 Glue and nail the roof panel in place, overhanging by ½ in. (12 mm) at each side, ¼ in. (6 mm) at the back, and 1½ in. (4 cm) at the front. Use 1¼-in. (30-mm) nails for this step, as the fence board is thicker than that used in the rest of the project.

8 Switch back to 1-in. (25-mm) nails and glue and nail the bark stopper to the floor, butting it up against the inside of the front.

9 Paint the outside of the butterfly house gray, let it dry, and then spray white on top. Let dry.

10 Apply silicone to the underside of the bark, then place it on the roof and screw in place with 1¼ in. (30-mm) exterior screws. (You may need longer or shorter screws depending on how thick the bark is.) Add driftwood pieces to the front by drilling pilot holes and then driving in 1-in. (25-mm) loose finish nails. Attach moss randomly on top of the bark and driftwood, using silicone to hold it in place.

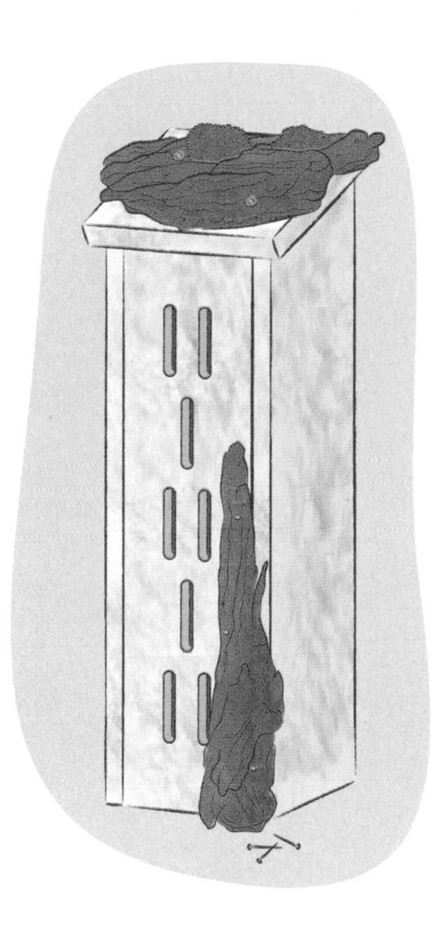

11 Attach a latch to the bottom of the lower back panel to keep the bark and sticks from pressing the door open. Keep it secured with a small lock or wire.

12 Place a couple of slabs of tree bark vertically inside the house, making sure they don't block the hinged lower back panel. This gives the butterflies more shelter and gives them something to cling on to.

Ladybug

ladybug house

Invite ladybugs (ladybirds) into your garden to eat aphids and other bad bugs. If you have plenty for these armored beauties to eat, they will inhabit and enhance your garden.

materials

One 6ft x 5½ in. x ⅝ in. (180 x 14 x 1.5 cm) cedar dog-ear fence board
6-ft (180-cm) length of 2 x 4-in. (5 x 10-cm) timber
Two 6-ft (190-cm) lengths of 2 x 1-in. (5 x 2.5-cm) weathered fencing
Waterproof premium glue
Wood putty
80-grit sandpaper
1-in. (25-mm) and 1¼ in. (30-mm) finish nails or galvanized wire nails
Two 1¾-in. (45-mm) loose finish nails (for hinged door)
Twelve 2½-in. (60-mm) exterior screws
Red, brown, gray, and white exterior spray paint
White, yellow, and green craft paint
Ashland® natural raffia
Approx. 100 flexible drinking straws, about ¼ in. (6 mm) in diameter
Approx. 100 x 12-in. (30-cm) bamboo skewers
Seven ladybug stick pins (available from craft stores)—optional
Weatherproof 30-minute clear silicone
Basic tool kit (see page 126)
Wire cutters

finished size

Approx. 21 x 8½ x 7¼ in. (53 x 21.5 x 18.5 cm)

cutting list

Left side panel: 8 x 5½ in. (20 x 14 cm) —cut 1, beveling the cut at a 30° angle
Right side panel: 6 x 5½ in. (15 x 14 cm) —cut 1, measuring 6 in. (15 cm) down the fence board from the point of the left side panel beveled edge and then making a straight cut
Base: 4½ x 5½ in. (11.5 x 14 cm)—cut 1
Back: 7⅛ x 4½ in. (18 x 11.5 cm)—cut 1, mitered at a 30° angle
Partition: approx. 4½ x 5½ in. (11.5 x 14 cm)—cut 1 (see step 4—measure the partially assembled box at this stage before cutting the partition panel)
Roof: 8 x 5½ in. (20 x 14 cm)—cut 1
Door: 4½ x 2 in. (11.5 x 5 cm)—cut 1

1 Glue and nail the left side panel to the base, making sure that the highest point of the beveled edge is on the outside.

2 Dry fit the back panel on top of the base to ensure that you don't need to cut a bit more off. Glue and nail it in place.

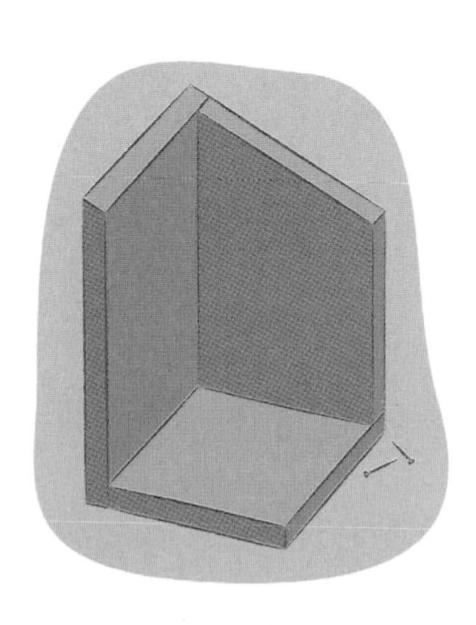

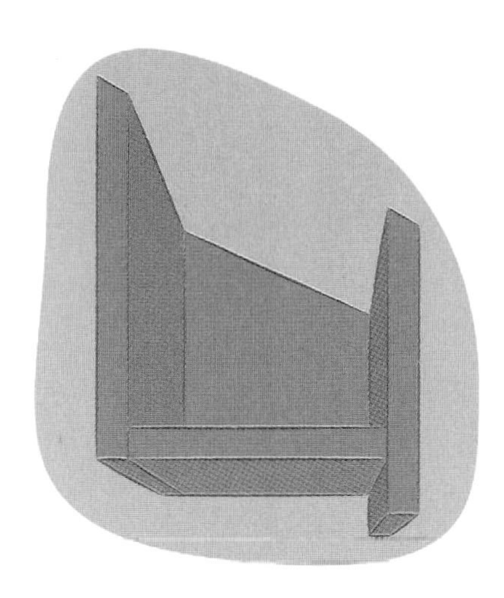

3 Lay the piece down, with the back panel flat on the table. Align the beveled peak of the right side panel with the back-panel edge, leaving a bit of the side panel hanging below the base. The roof will be attached to the open, slanted part of the box.

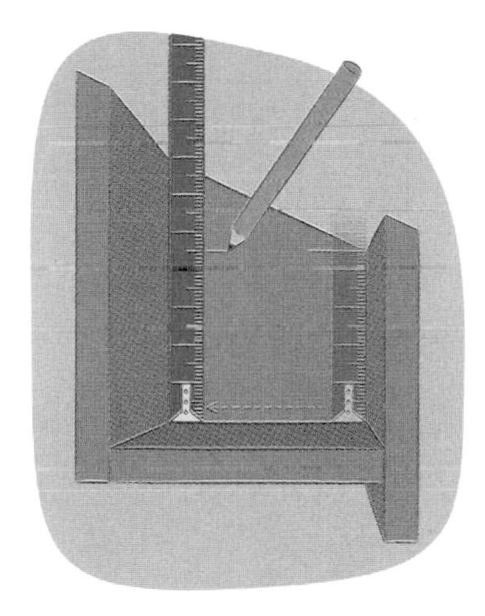

4 Measure the shortest depth of the box to find where the partition needs to go; it will be approx. 4½ in. (11.5 cm). (This is where the straws and bamboo skewers will rest.) Mark this measurement on opposite sides of the box. Cut the partition panel to size. Apply glue to the inside of the box between the marked points, slide the partition in, and nail it in place.

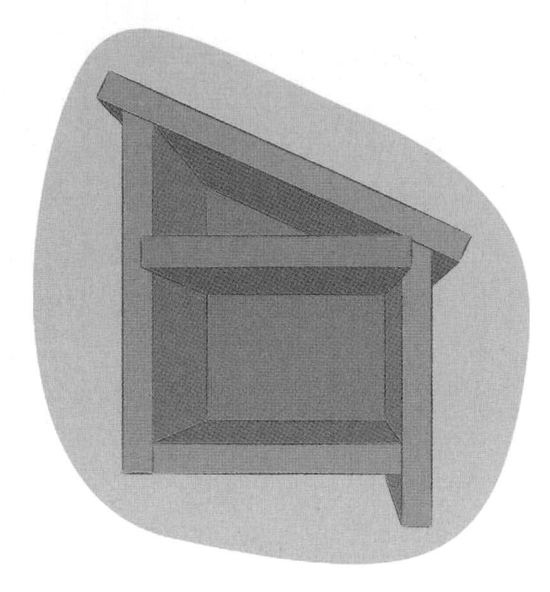

5 Apply glue to the top edges of the back and side panels, then place the roof panel on top, with ½ in. (12 mm) overhanging at each end. Nail in place.

6 Paint the bug house and door in your chosen color(s)—I used red for the house and door, then undercoated the roof brown—let dry, and then dry brush white on the roof.

7 Fill the bottom portion of the bughouse with natural raffia—great bedding for ladybugs. Cut about 100 straws roughly 4½ in. (11.5 cm) long (the measurement doesn't have to be exact) and fill the top part of the bughouse, interspersing bamboo skewers from time to time to make "landing pads" for the ladybugs.

Use wire cutters to cut the bamboo skewers to random lengths; scissors won't work.

8 Place the door ½ in. below the bottom of the partition. Measure 3¼ in. (8 cm) up from the bottom of the base on each side of the bughouse. Using a 1⁄16-in. (1.5-mm) bit, drill a pilot hole (see page 131) in the door at these points, making sure they are level. Drive in loose finish nails to hinge the door in place.

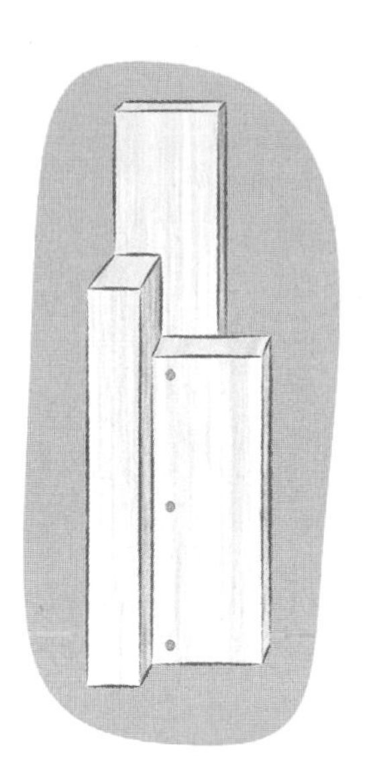

9 Now make the base stand. From 2 x 4-in. (5 x 10-cm) timber, cut three pieces 12, 15, and 19 in. (30, 38, and 48 cm) in length. Place them together as shown and screw together with 2½-in. (60-mm) exterior screws, making sure that the bottom stays level. Paint gray and let dry, then drybrush white paint over the top to give a weathered look.

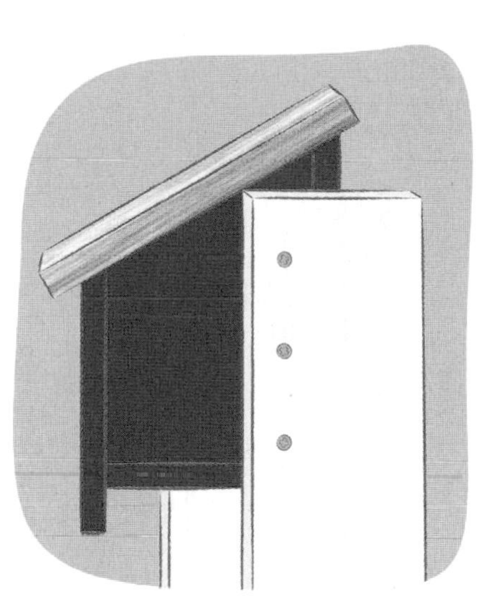

10 Place the bughouse on top of the shortest 2 x 4-in. (5 x 10-cm) post, pressed up firmly against the other posts. Drill pilot holes in the tallest post, then drive in 2½-in. (60-mm) screws to attach the bughouse to the post.

11 Cut 2 x 1-in. (5 x 2.5-cm) weathered fencing into random lengths ranging from 3 to 10 in. (8 to 25 cm). Glue and nail them around the bottom of the base stand, using 1¼-in. (30-mm) nails. If you wish, add ladybug stick pins by applying silicone to the base and gently hammering them in place.

Tools and materials

Although there are undoubtedly some useful power tools and cool gadgets that will make things easier for you, the houses and feeders in this book can all be made with a fairly basic range of woodworking tools and equipment.

Basic tool kit

The following tools are what I consider to be the basic requirements. Any extra tools are included in the materials list for each project.

- Miter box
- Hand saw
- Electric chop miter saw
- Power drill
- Drill bits: 1/16-, 1/8-, 3/16-, and 5/16-in. (1.5-, 3-, 5-, and 8- mm)
- Hole saw bits: 1/8, 7/16, 5/8, 11/16-, 7/8-, 1, 1 1/8-, 1¼, 1½ -, and 2-in. (3-, 11-, 15-, 17-, 22-, 25-, 28-, 30-, 38-, 42-, and 50-mm)
- Hammer
- Finish nail gun
- Screwdrivers: Phillips head, straight edge
- Metal tape measure
- Speed square and/or set square
- Vice and clamps
- Old rags/T-shirts
- Paintbrushes ranging in size from ¼ in. to 2 in. (6 mm to 5 cm)
- Putty knife
- Pencil
- Sanding sponge

safety first

These rules apply when using saws or power tools for cutting wood, glass, etc.

- Keep your work area clean to avoid trip hazards.
- Always push up loose-fitting sleeves, tie up long hair, and don't wear long necklaces or scarves.
- Wear safety glasses to protect your eyes from sawdust and splinters, a dust mask, and ear protectors to preserve your hearing.
- Always clamp your wood in place so that it doesn't move while cutting.
- Always unplug a power saw before changing the blade.
- Let the blade come to a complete stop at the end of a cut before lifting the saw.
- Never reach under the saw when the blade is rotating, even if it has a guard.
- Guards should never be removed or forced open.
- Always follow the manufacturer's instructions and safety guidelines.

Wood

Cedar dog-ear fence board

I use cedar dog-ear fence board for my birdhouses and feeders, for several reasons. First, it's relatively inexpensive. Second, it is actually better for the birds, since it is raw, untreated, and without chemicals that will harm these wonderful creatures, which means that the birds are not exposed to toxic fumes from varnish or paint inside the nest box! Fence wood is rough, which means that it's easier for birds' little claws to grab hold of. And finally, standard fence boards are a good width for the birdhouse panels (5½ and 7½ in./14 and 19 cm)—so you can simply cut across the board without having to take lots of unwanted measurements and markings.

Cedar is a soft wood—easy to cut, but also easy to warp. Lay it flat with ties to prevent it from warping. Dry heat plays a big factor in wood warping, splitting, bowing up, or shrinking, as well. If your area is dry and hot, I suggest getting what you need for your project, instead of storing several pieces. Otherwise you will be throwing away wood that's only good as firewood.

Never use plywood for a birdhouse, as it will warp and fall apart. Smooth wood that is used for building furniture isn't a good idea either, since it is not rough enough and then you will have to cut grooves in the interior, so the birds can use their tiny claws to get out. Pressboard should definitely not be used for birdhouses: it will disintegrate in bad weather, and it falls apart once you drill screws through it.

knots in wood

Cedar wood boards have knots, which can add visual interest to your project, but be aware of where they are. If you've got a knot along the edge, avoid it when nailing, especially with a powered nail gun. Knots are hard, which can make a nail turn and cause injury; knots can also split when a nail is shot directly into it, thus ruining your project. Unless you're planning on popping out the knot itself to make an entrance hole, avoid placing the knot near where a hole is to be bored with a hole saw bit.

tips

- When handling cedar fence boards, wear gloves if possible to avoid getting splinters in your hands or fingers.

- When sanding edges, wrap the sandpaper around a block of wood or sanding sponge to avoid wood splintering into your hand or into your fingernail. You can buy sanding sponges from the paint section of your home improvement store: simply cut your sandpaper in half and wrap it around the sanding sponge. It's soft and flexible, making it easier to work with.

- Buy the driest boards if you are starting your project right away. Check for damage on edges, splitting, and an excessive number of knots. If the wood is too wet, it will cause your glue to lose its integrity and not bond well. Warping will also be a big factor if the wood is wet and you do not store it properly.

1 x 1-in. (25 x 25-mm) square wood dowel
I sometimes use this for roof ridges (see the Hip Bird Feeder on page 85). You can also use square wood dowels as door knobs or as a guardrail on the birdhouse front porch.

Glues, paint, and varnishes

As your birdhouses and feeders will be outside and therefore exposed to the elements, it's important that they're as weatherproof as they can be. Be sure to use waterproof premium glue, so that the pieces are firmly adhered together. You can also use weatherproof 30-minute clear silicone to adhere decorative items such as moss, buttons, glass, etc.

Use craft paint or spray paints to decorate the exterior of birdhouses, and varnish them with exterior varnish. Never paint or varnish the interior of a birdhouse, as this could be very harmful to the birds. Before you paint or varnish, always make sure that the entrance hole is blocked with wadded-up paper or tape, so that nothing will accidentally drift inside. If paint seeps through onto the inside of the hole, simply use sandpaper to remove it. For some very simple but effective decorative paint techniques, turn to page 139.

Techniques and guidance

Throughout this book you'll learn new and old techniques that many woodworkers have used for years and years. There are also a few techniques that I have learned from my own experiences and experimentation that I'm sharing within this book. Most of the projects are simple with miter and angled cuts.

Cutting

In most of the projects in this book, the full width of the dog-ear fence boards is used to create the panels, so that cutting is reduced to the minimum. In addition to straight cuts, you also need to know how to make miter cuts (angled cuts, usually at 45°, so that two adjoining pieces will fit together in a tight right angle), and bevel cuts, which means cutting an edge on a slant. You can cut most of the pieces for the projects in this book using a simple miter box and backsaw; the only exceptions are those that involve 22.5° bevel cuts, for which you will need a compound miter saw. The steps below show how to cut with different tools.

grain of wood

Fence wood is milled with the grain of the tree (lengthwise). When cutting fence wood, lay the board lengthwise with the edge against the fence of your miter box or miter saw. Cut across the grain at the length specified. The grain of wood will run in the long direction of the front, back, and sides of panels.

Traditional wooden miter box

Use a traditional wooden miter box and a backsaw for making simple miter cuts. The miter box has slots at 45° (sloping right and sloping left) and 90° into which the backsaw fits, so that you can cut at an accurate angle.

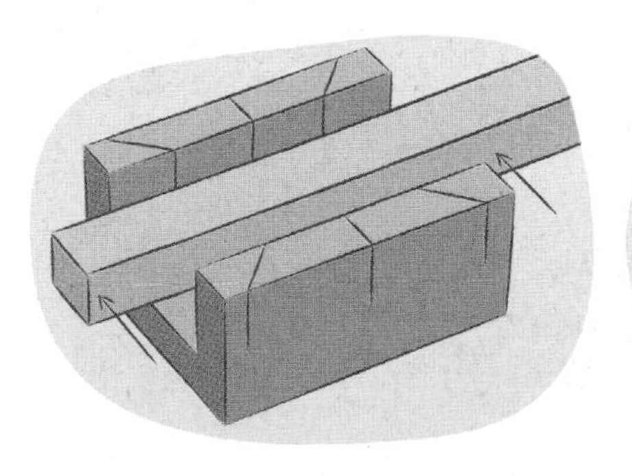

1 Lay the piece you're cutting in the miter box and hold it tight against the back fence.

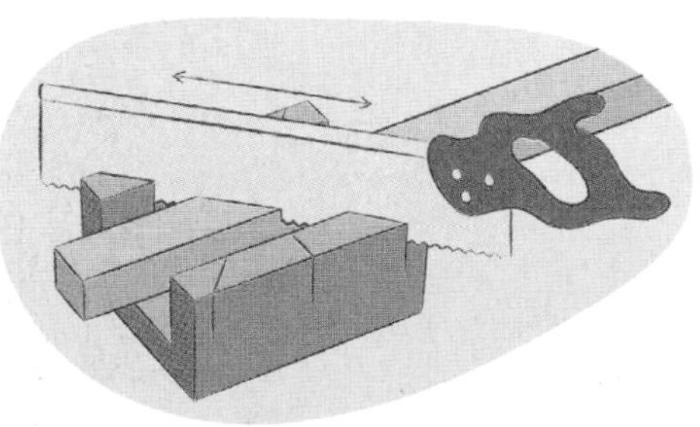

2 Set the backsaw in the appropriate slot of the miter box and make the cut, using slow, smooth strokes, letting the blade do the work. Never force your blade to cut the wood: doing so can result in serious injury.

Adjustable miter box

There are a number of adjustable miter boxes on the market that allow you to cut miters at any angle from straight cut up to 45°.

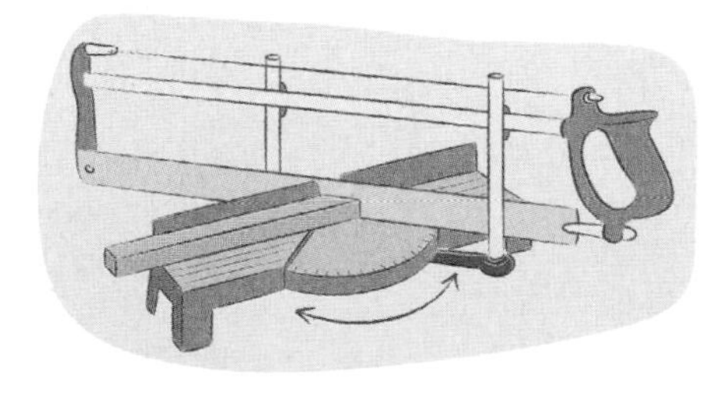

1 Rotate the saw to the desired angle, lock it in place, then make the cut.

Compound-angle power miter saw

This kind of power saw allows you to make miter cuts and also to bevel the edge of the piece; moreover, you can set both the miter angle and the bevel angle at the same time. As you might expect, this tool is more expensive than the others shown here, but if you do a lot of woodworking, it is well worth buying one for its ease of use and the amount of time it will save you.

Miter cut

To make a miter cut, simply rotate the saw blade to the right or left to the appropriate angle and lock it in place.

Bevel cut

To make a bevel cut, tilt the blade to the appropriate angle and lock it in place; note that some saws only tilt to the left, but dual-bevel models tilt both to the left and to the right.

Drilling entrance holes

When configuring where your entrance hole is going to be, you first need to know which species of bird you want to attract. Chickadees, for example, like 10–12 in. (25–30 cm) from the entrance hole to the floor. The chart on page 140 sets out entrance hole sizes for different species.

The entrance hole is drilled before the birdhouse is assembled. I usually add ⅝ in. (1.5 cm) to my measurement to take into account the bottom floor panel.

Hole saws are the best way to make a perfect entrance hole for birdhouses and they easily attach into your drill.

When drilling a hole into the front panel, lay it on a solid surface, such as a cutting board. I place my wood piece on a kitchen cutting board on the floor of my workshop and hold it in place with my closed-toed boots. This avoids ruining my work surface or dulling the drill bit.

Joining

Joining pieces together is easy when you use the proper tools. I find that having a vice grip on hand helps when gluing and nailing pieces together. Clamps are a must when your wood decides to warp, and you can clamp pieces together while nailing.

Gluing and nailing

I always use waterproof premium glue and galvanized nails for extra security. You can screw pieces together with exterior screws if you wish to make them even more sturdy—just remember to drill pilot holes (see opposite) for the screws.

1 Apply a small amount of waterproof premium glue along the center of the two edges that you are joining together.

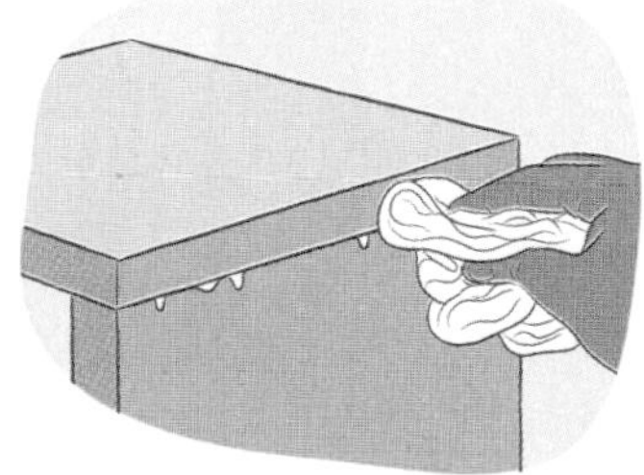

2 Press the edges together, then wipe off any glue that oozes out of the join with an old rag, a wet sponge, or an old paint brush dipped in water. Clean glue off with water.

3 Using a nail gun, drive 1-in. (25-mm) finish nails through both pieces, making sure your fingers are not anywhere near where the nail could turn and come out, hitting your finger.

Drilling pilot holes and driving in screws

A pilot hole is a small hole that you drill before driving a screw into a piece of wood. It prevents the screw from splitting the wood and it ensures that the screw will be installed straight, because it will follow the path of the pilot hole. Remember that the bit end gets extremely hot from the friction and can cause severe burns on fingers or hands.

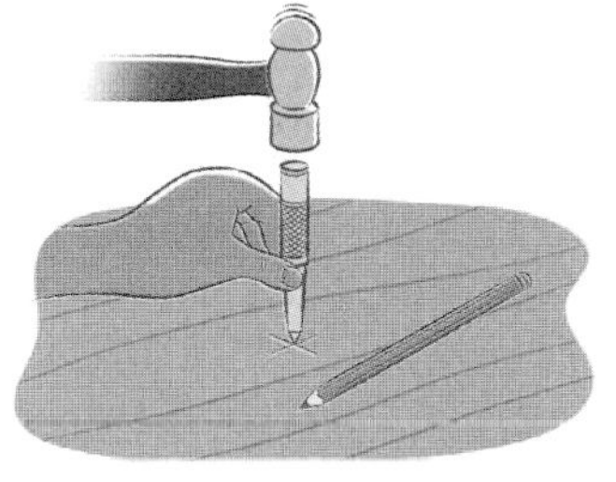

1 Using a pencil, mark on the wood where you want the screw to go. Make a small indentation in the wood using a center punch—a small, slender tool with a pointed end. This will help stop the drill bit from slipping when you start the pilot hole. Position the tip of the punch over your pencil mark, then strike the punch gently with a hammer. (For very small screws, you can use a bradawl instead of a center punch.)

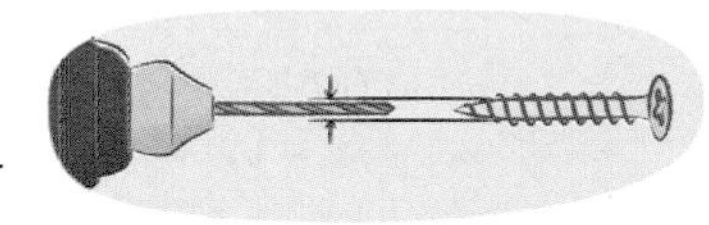

2 Insert the appropriate bit into your drill: the bit for a pilot hole should be smaller than the diameter of the screw you're intending to use and, for the projects in this book. A 1/8-in. (3-mm) bit will usually do the job. (Remember that you can always make the pilot hole bigger in relation to the size of your intended screw.

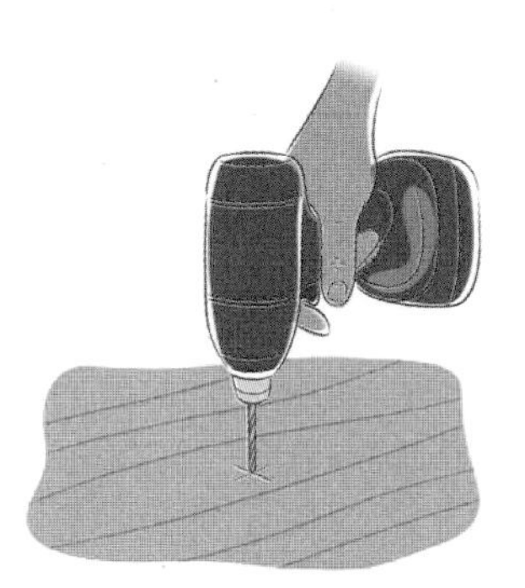

3 Place the tip of the bit in the indentation you made in step 1, angling it at the angle you want the screw to follow, then drill the hole to a depth equal to the length of the screw. Back the bit out carefully.

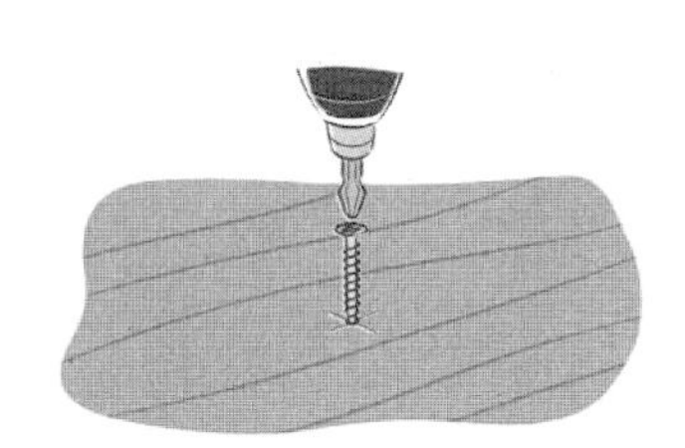

4 Fit your drill with a screwdriver bit. Place the tip of the screw in the pilot hole and drive the screw in, angling it so that it follows the path of the pilot hole. Don't overtighten your screw, because this will split the wood.

Countersinking screws

Countersinking screws simply means driving them in so that they sit flush with the surface of the wood. There is a special bit for doing this. The bit makes an indention in the wood above your pilot hole, then the screw head fits snugly, flush with the membrane of the wood. You shouldn't drill the screw head below the membrane, as doing so will cause water to pool and could crack your wood.

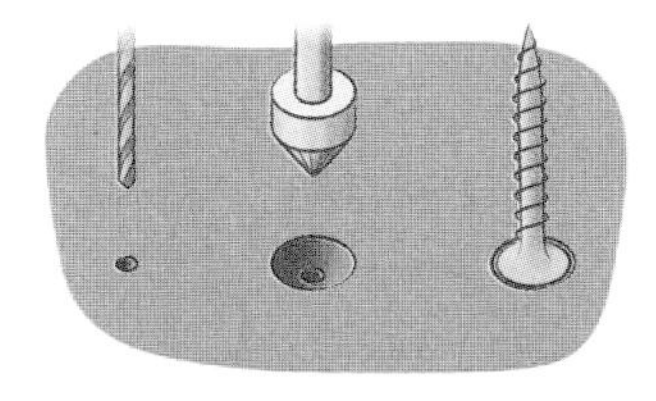

1 Drill a pilot hole for the screw (see above). Using a countersink drill bit with a multiple cutting edge, begin making the screw hole. Go down a little way, then remove the drill bit. Place the head of the screw in the hole you've just made to check the size: the hole should be fractionally larger than the screw head.

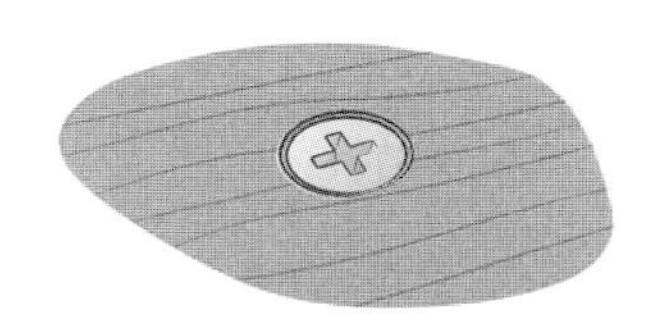

2 Insert the screw until the head is flush with the surface of the wood. Fill with wood filler putty, leave to dry, then sand the filler flush with the surface of the wood.

Tip

Fill nail and screw holes with glue, and then with paintable wood putty. The glue will help keep the nails from moisture and might help prevent popping from endless expansion and contractions due to changing weather conditions, while the putty will give you a smooth surface for painting.

Hanging houses and feeders

Hanging birdhouses can be hung from an iron plant hanger, tree limb, or from a hook on your eaves where predators are less likely to reach them. I generally use pipe clamps (clips), which are normally used to clamp a water or electrical pipe to a wall and can be bought from the plumbing department of home-improvement stores, as a base for hanging the birdhouse. You can also use tie wire strung through the roof or through driftwood limbs, which creates a more natural look.

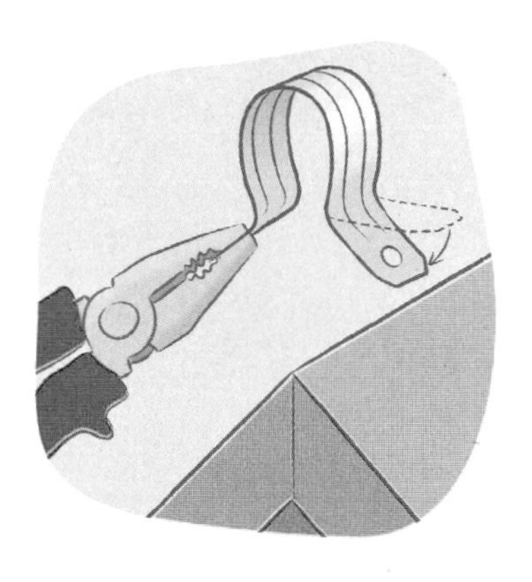

1 Bend a 1-in. (25-mm) two-hole pipe clamp (clip) with pliers to fit on each side of the roof. Mark the center of the apex of the roof. Place the pipe clamp (clip) over the apex at the center point and make a small pencil mark through the center of each hole.

2 Using a 1⁄16-in. (1.5-mm) bit, start a hole at each marked point. Screw the pipe clamp (clip) in place with ¾-in. (20-mm) exterior wood screws, being careful not to overtighten.

3 You can then either hang the pipe clamp (clip) from a metal hook in your garden, or loop wire through the fastener and hang the birdhouse from the wire.

Wire

Another option is to attach wire to the back of projects such as bee houses and/or ladybug houses. This method is beneficial if the project is going to be placed flush against a wall or hung on a fence or post.

1 On the back, drill two pilot holes using a ⅛-in. (3-mm) bit, and insert a 1¼-in. (30-mm) screw partway into each one. Cut a length of 18-gauge (1-mm) wire, wrap each end around one of the screws, then drive the screws in almost flush to form a hanging loop, being careful not to overtighten them, as this may crack the wood.

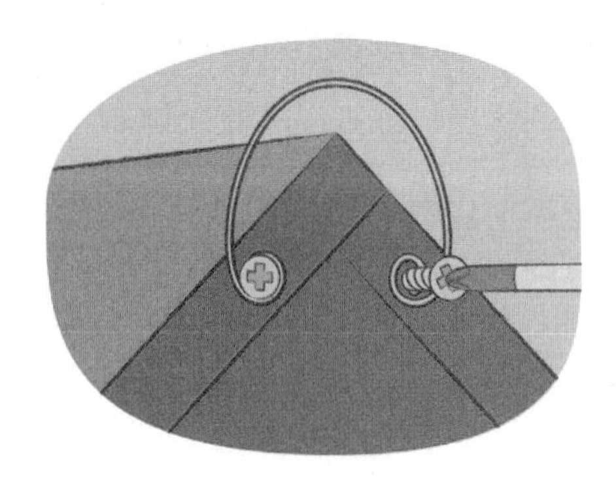

Chain

Alternatively, you can attach screw eyes to the roof of the birdhouse or feeder and then attach single-jack chain—just make sure the chain will take the weight of the project. The project instructions will tell you where to attach the screw eyes and how many lengths of chain are required. I either attach four lengths—as shown below—or two lengths to diagonally opposite corners and join them all together in the center.

1 Insert screw eyes into the roof of the birdhouse or feeder, following the instructions in the project.

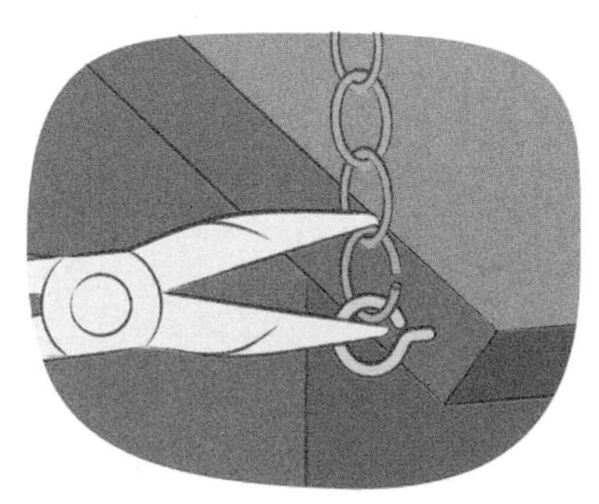

2 Cut the chain to the required length(s). Using needle-nose pliers, open the first link in each length of chain. Loop each one through a screw eye and close it again.

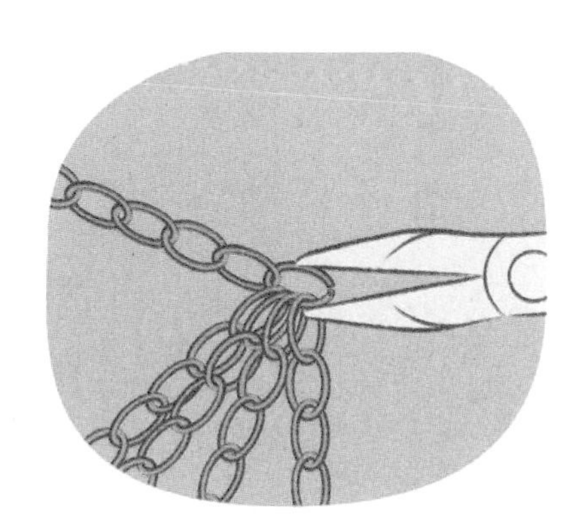

3 Finally, cut and open up one link from the remaining chain, loop it through the unattached ends of all four chain lengths, and close again to join all four lengths together ready for hanging.

Post-mounted projects

Mounting a birdhouse or feeder on a galvanized steel post is a great way to raise it off the ground and protect the birds from predators such as cats, possums, snakes, squirrels, and other feather-eating monsters. The method shown here uses items from the plumbing and fencing departments of your local home improvement store. Pipe flange can be found in the plumbing section. Rebar is found in the concrete section and is normally used in strengthening concrete.

1 Mark the center of the base of the birdhouse. Place the pipe flange hole over your pencil mark. Make pencil marks where the screws are needed on the pipe flange. Using a 1/8-in. (3-mm) bit, drill pilot holes at the screw points. Use ¾–1 in. (20–30-mm) screws to ensure they do not penetrate the inside of the birdhouse; the screws that come with the pipe flange may be too long.

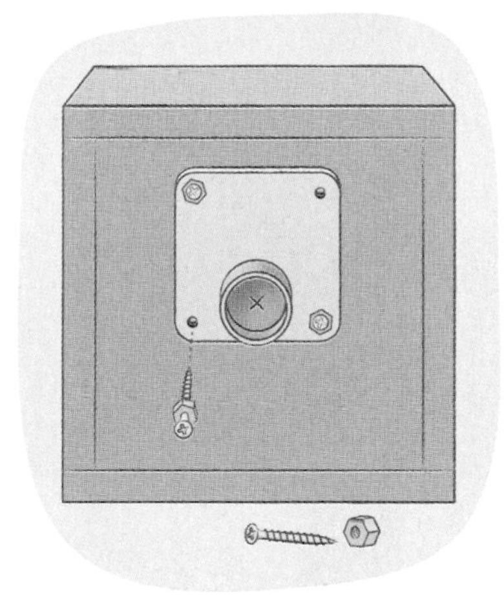

2 Thread the appropriate pipe onto the pipe flange, then slide the pipe into the hollow opening of galvanized fence pipe.

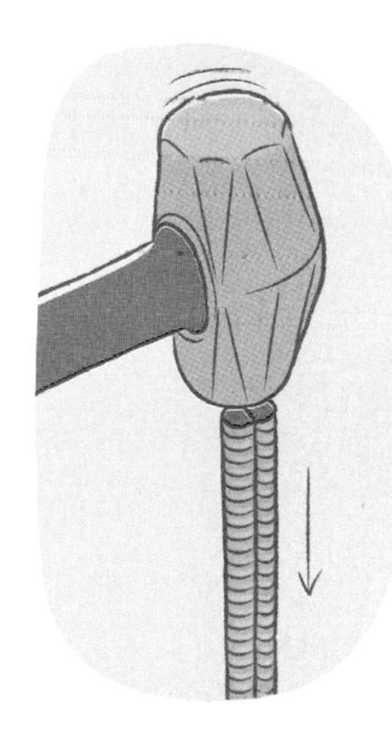

3 Decide where you are going to position your project. Using a sledgehammer, pound two 4-foot (1.2-m) rebar/reinforcing bars—or the appropriate length—into the ground to roughly 12–18 in. (30–45 cm), making sure they are secure. Alternatively, you can set the rebars in concrete: dig a 12-in. (30-cm) hole, place the pole in the center, pour in a bag of quick-setting concrete and soak with water (follow the manufacturer's instructions), checking the pole is level. When the concrete has set, slide the galvanized fence pipe over the rebars.

4 To hold the project firmly in place, drill pilot holes through the fence pipe from opposite sides and insert bolts, tightening them against the sides of the rebars with a ratchet drill bit (optional).

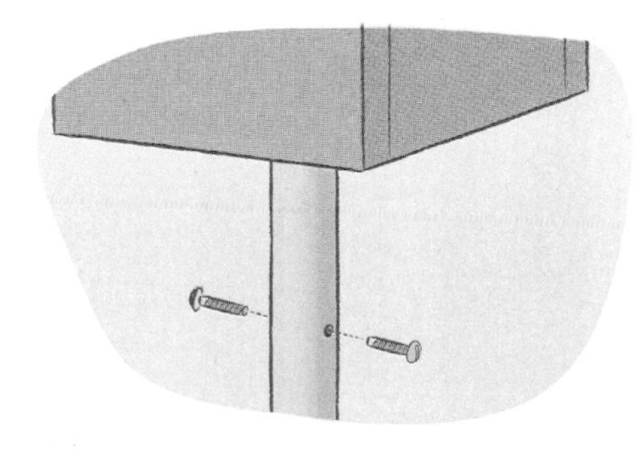

Bird safety

We all want to encourage birds to nest in our gardens, but it's absolutely vital to take some basic safety precautions so as not to injure your feathered friends.

- Make your birdhouses from untreated, unvarnished timber, as wood preservatives and varnishes are harmful, if not fatal, to birds. I use untreated dog-ear fence board, because it's the perfect size for birdhouses for small cavity-nesting birds. The boards are left rough on both sides, so the birds can grab onto it with their tiny little feet and claws for exiting the birdhouse. This is especially important for the young fledglings once they're ready to leave the nest.

- Good ventilation in your birdhouse is the best way to keep the inside cool. The roofs on the birdhouses in this book are built to give plenty of circulation throughout. If there is a different type of roof, like the simple roof (page 9), the ventilation holes are normally drilled on each side of the birdhouse just under the eaves, sloping downward, so that when it rains the water will drip away and not run down inside the box.

- If you are concerned about rain entering into the birdhouse, simply drill 1/8-in. (3-mm) holes in the bottom. This will allow water to drain out.

- Many people ask me if the birds will fit through the tiny entrance holes. The answer is yes. Birds that will go into birdboxes generally require a small hole. Although there are some species that need larger holes and boxes, for the most part, these tiny cavity-nesting birds can fit holes 1 1/8–1¼ in. (28–32 mm) in diameter. I have yet to see a bird turn down a house because of the specifications. My saying is; "If it fits, it's home, tweet home!"

- When you paint your birdhouse, make sure that you block up the entrance hole with wadded-up paper or tape, so that no paint can accidentally stray into the interior. Let the paint dry for a few days before you put the birdhouse out in the garden for the birds to investigate. This will ensure that any fumes or toxins will be well cured and not distract birds from entering your lovely birdhouse, as the harmful odors will have dissipated by then.

- Adding a dowel as a perch is purely decorative. Birds have sharp claws, which makes it easy for them to hang on to the rough exterior of the house and check it out. But remember that the dowel or any objects attached to the front of the birdhouse will also give predators something to hang on to when searching out the fledglings. Unfortunately, that's the nature of the beast.

- Think carefully about where you position your birdhouse. Getting a little sun is good, as long as it's not for a long period of time. Take the time to stand in the place you're planning to put your birdhouse and see how warm the sun is on your face or body. If it's uncomfortable after just a few minutes, think of how it would be if you were inside the bird box.

- If you are going to hang your birdhouse from a tree, try placing it on a branch that's away from the main trunk, where predators won't be tempted to disrupt the birdhouse. The best option (I feel) is to mount the birdhouse on a pipe post 6 ft (1.8 m) or so off the ground, as the post is slippery, preventing predators from climbing up—but this may need some shade from the hot afternoon sun, depending on where you live.

- Encourage the birds into your garden by providing them with material that they can use to line their nests; see the hanging material basket on page 98. Near the birdhouse, you can leave a ball of yarn, Spanish moss, twigs, and other stringy objects they will use to build a nest. You can also encourage birds by planting flowers and bushes that provide fruit and seeds.

- This book features a range of imaginative ideas for bird feeders. However, you should never try to attract the birds to your birdhouse by placing food in or on it, as other critters that like to eat bird seed, such as squirrels, rats, chipmunks, and even bears, will also be attracted, thereby destroying the birdhouse and its occupants. Bird feeders are for the food and birdhouses are for nesting and rearing the youngsters.

- A simple tip for creating the perfect bird garden is to place each component (feeder, water, and birdhouse) in a triangle at least 15–25 ft (4.5–7.5 m) apart.

Birdhouse maintenance

With a little care and simple maintenance, your birdhouse should last for years and you will find that birds will return time and time again to nest and bring up their young. I have occupants in my birdhouses every year and love every minute of it.

Once the fledglings have left the nest, remove the nest with a putty knife and clean the inside of the birdhouse so that it's ready for the next season. With the exception of the airplane birdhouse (page 58), all the birdhouses in this book have clean-out doors that you can open for easy nest removal. Wear gloves when you do this to ensure you're not infecting yourself with parasites or diseases that may be lingering in the nest. Birds will return to their nest each year, but it's best to remove the nest each season and wash out the interior with soap and water (never use bleach or harsh chemicals) to ensure that bacteria doesn't grow, which can cause illness to the new hatchlings. Let the interior of the birdhouse completely dry out, then return it to its original place.

The birdhouses in this book have been painted with exterior paint. However, as with all exterior materials, care is required. After you've removed the nest and cleaned properly, varnish the exterior of the birdhouse, let dry completely for a few days, then return it to its original place. Be sure to keep the entrance hole clean of varnish. Do this each season a few months before the birds start nesting. If you have extremely harsh weather, bring your birdhouse indoors or set it on a covered porch.

Decorating your houses and feeders

Your birdhouse can be whimsical, rustic, modern, or just out-of-this-world eclectic. Use your imagination when embellishing your projects.

Decorative items

I pretty much use anything and everything to decorate my projects, from scrapbooking gadgets to old garden tools, as long as I can secure them down to withstand the elements. In search of items, I walk the aisles of craft stores, home improvement stores, antique shops, flea markets, and even rummage through...yep...trash! (I'm not talking garbage—just what people throw out that can be taken apart and re-cycled. For example, I had some shoes I was going to give to the Goodwill that had charms on them. Instead, I removed the charms and used them for a project.) Parts from furniture, such as old knobs and hinges, also work well.

Driftwood can be found by walking the shore of the ocean, creeks, or just out in the wilderness. Just take care to clean it well and make sure there are no pests lurking in the crevices. Washing with soap and water is a good rule to follow with any objects picked up from the outdoors.

In order to keep birds safe, please take care so as to not get any harmful chemicals inside the birdhouses. Always place tape around the inside edges of the entrance hole and stuff with wadded tissue before painting. The bugs don't eat or peck on the wood, so this is not as important for them as it is for our feathered friends.

The "weathered" look

This process is a great way to make a project look old and weathered or, as they say "shabby chic."

Paint a base coat of your chosen color, let it dry completely, then choose a color that will contrast with the first. Apply your second color choice over the first and, while it is still wet, drag a piece of torn cardboard down and from side to side, exposing the base coat. You will need several torn pieces of cardboard in order to do this process properly. Let each side dry before you start the next.

You can also use a 1-in. (2.5-cm) dry brush to pick up a small amount of the second paint and, using soft, swift strokes, apply it over the first coat.

Another way to create this look is to spray water directly onto the painted (already dry) surface, then spray paint in your chosen color over the surface while the water drips down. This is very messy and the paint will drip onto your surface. Hang your project from a wire over a trash can and let the excess paint drip inside the trash can.

Hand-painted flowers

In this simple technique, you apply two colors at once to create simple but natural-looking flowers.

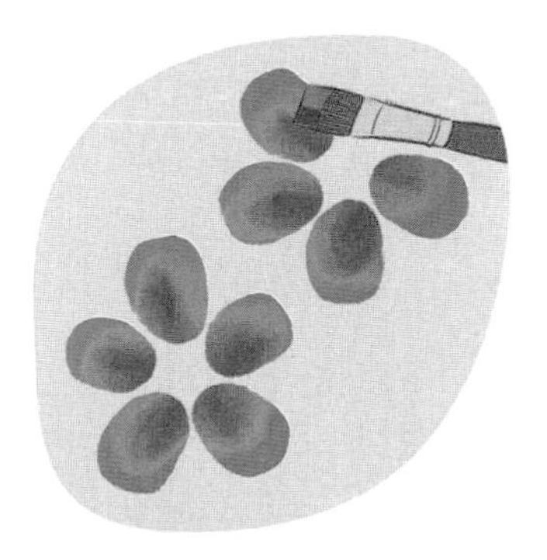

1 Place two drops of paint color on a clean surface next to each other. Dip an angled 1-in. (2.5-cm) brush into both colors at the same time, loading the brush. With a single stroke, press the brush in a downward motion (practice on a piece of clean cardstock or paper), then lift.

2 Continue this process in a circular motion to create a complete flower or you can use these strokes from side to side to create the side view of a flower.

Birdhouse sizes for different species

Many species of cavity-nesting birds will nest in man-made boxes but, just like humans who are looking for a new home, they have their own preferences and requirements. It's important to know what type(s) of birds you're building the birdhouse for. You'll have greater success attracting the birds you want to your birdhouses if you build them to the proper size.

The entrance hole size probably is the most critical factor when building birdhouses. If it's too small, your chosen bird species may not be able to enter the house. If it's too large, it could allow bigger, more aggressive species to enter. Some species prefer to nest in an open-sided box with the top half of one side of the box removed.

Some tips on siting boxes

- Allow the birds a clear flight path to the entrance.
- Place the box in a higher, more inaccessible place when there are predators such as cats around; if possible, give the box protection through siting it above a thorny bush.
- Ensure that the box is suitably sheltered from rain, so that water will not flood in through the entrance.
- Try to have your nest box in place by January or February, as birds will start prospecting for nest-sites early if there is a spell of warm weather (sometimes even the previous fall/autumn!). However, birds will use them at any time of year for roosting, so it is worth putting them up as soon as they are ready.
- Keep nestboxes away from bird feeders, as otherwise the occupants will constantly be fending off intruders that come to feed on their doorstep.
- The chart on the right shows some of the most popular cavity-nesting birds of Europe and the United States. Check which ones visit your garden, and then design your birdhouse accordingly.

Black-capped Chickadee, Mountain Chickadee, Boreal Chickadee
The Black-capped Chickadee is well known for its mating call of "chick-a-dee-dee-dee, dee-dee"; these birds are a favorite of many who love watching birds in their gardens.

Distribution: All are resident in northern North America: Black-capped in northern half of United States and most of Canada; Mountain in the Rocky Mountains; and Boreal in most of Canada and the extreme northern parts of the United States.

Entrance hole: 1⅛–1½ in. (28–32 mm)
Entrance hole to floor: 5 in. (12.5 cm)
Floor size: 4 x 4 in. (10 x 10 cm)
Box height: 8 in. (20 cm)
Height above ground: 6½–16½ ft (2–5 m)

Bewick's Wren
These master vocalists belt out a string of short whistles, warbles, burrs, and trills to attract mates and defend their territory, or scold visitors with raspy calls. Bewick's Wrens favor brushy areas, scrub and thickets in open country, or open woodland. These birds normally breed in areas that contain a mixture of thick scrubby vegetation and open woodland

Distribution: Bewick's Wrens are still fairly common in much of western North America, but they have virtually disappeared from the East.

Entrance hole: 1–1¼ in. (25–31 mm)
Entrance hole from floor: 4–6 in. (10–15 cm)
Floor size: 4 x 4 in. (10 x 10 cm)
Box height: 6–8 in. (15–20 cm)
Height above ground: 6–10 ft (1.8–3 m)

Blue Tit, Coal Tit, Marsh Tit, Willow Tit
The Blue Tit is a common visitor to gardens, even in towns and cities, while the other three species are more tied to woodlands and large parks.

Distribution: All species are widespread residents across most of Europe; the range of Coal and Willow Tits extends across northern and central Asia.

Entrance hole: 1 in. (25 mm)
Entrance hole to floor: 5 in. (12.5 cm)
Floor size: 4 x 4 in. (10 x 10 cm)
Box height: 8 in. (20 cm)
Height above ground: 6½–16½ ft (2–5 m)

Eastern Screech Owl
Screech owls live in forested parks and the edges of woodland clearings. They survive well in suburban neighborhoods with trees and use the spacious lawns as hunting grounds. These owls do not build any nest inside the box. Instead, the mother lays her 2–6 eggs on whatever is in the bottom of the cavity. Add 2–3 in. (5–7.5 cm) of wood shavings at the bottom of the nest box.

Distribution: Screech owls are year-round residents in nearly every state across the US. They primarily inhabit woodlands, but they are also commonly found in suburban and urban areas. They are one of the smallest owls in North America, standing about 10 in. (25 cm) tall with a wingspan up to 24 in. (60 cm).

Entrance hole: 3 in. (7.5 cm)
Entrance hole to floor: 9–12 in. (23–30 cm)
Floor size: 8 x 8 in. (20 x 20 cm)
Box height: 12–15 in. (30–38 cm)
Height above ground: 10–30 ft (3–9.1 m)

Great Tit, Crested Tit
These birds generally prefer to nest in a hole in a tree or wall, but a birdhouse mounted on the side of the house or on a tree trunk will do just fine.

Distribution: Both species are widespread residents in Europe, although in Britain the Crested Tit is restricted to the Caledonian pine forests of Scotland. The range of both species extends into Asia. The Great Tit is a common garden bird in all habitats, even cities.

Entrance hole: 1⅛ in. (28 mm)
Entrance hole to floor: 5 in. (12.5 cm)
Floor size: 4 x 4 in. (10 x 10 cm)
Box height: 8 in. (20 cm)
Height above ground: 6½–16½ ft (2–5 m)

Barn Swallow, Welcome Swallow, Greater Striped Swallow

Normally swallows are found near barns, nesting inside buildings on ledges and under eaves, and within such buildings they will utilize specially made platforms for nesting.

Distribution: Barn Swallow breeds in temperate regions of North America and Eurasia and winters in Central and South America, south Asia, and Africa. Welcome Swallow breeds in Australia and New Zealand; Greater Striped Swallow breeds in southern Africa.

Floor size: 5 x 5 in. (12.5 x 12.5 cm)
Box height: open-topped, 1½ in. (4 cm) high sides forming a shallow tray
Height above ground: 6½ –13 ft (2–4 m), situated against a wall or roof-beam.

Purple Martin

These birds nest in multiples of bird boxes that are specifically designed for the species.

Distribution: East and west North America; they winter in South America.

Entrance hole: 2 in. (50 mm)
Entrance hole to floor: 2 in. (50 mm)
Floor size: 6 x 6 in. (15 x 15 cm)
Box height: 6 in. (15 cm)
Height above ground: 10–17 ft (3–5 m)

Red-breasted Nuthatch, Pygmy Nuthatch, Eurasian Nuthatch

These birds are fun to watch as they climb the sides of trees, scampering upward from side to side, finding insects in the bark.

Distribution: Red-breasted is a widespread breeder across northern North America and in the Rockies; winters throughout the United States and southern Canada. Pygmy has a scattered distribution in the western United States. Eurasian is a widespread resident across temperate parts of Europe and Asia.

Entrance hole: 1½ in. (40 mm)
Entrance hole to floor: 6 in. (15 cm)
Floor size: 6 x 6 in. (15 x 15 cm)
Box height: 9 in (23 cm)
Height above ground: 10–17 ft (3–5 m)

Common Starling, Spotless Starling

Starlings can mimic a variety of songbird calls and tend to take over birdhouses occupied by other species. The Common Starling is now considered to be a pest in North America, but it is still favored by many people. The population is in decline in their native Europe.

Distribution: Natural breeding range is across Europe and into Asia; winters south to North Africa; introduced elsewhere around the world including North America, South Africa, Australia, and New Zealand. Common Starling does not breed in Iberia and North Africa, and here it is replaced by the very similar Spotless Starling.

Entrance hole: 2 in. (50 mm)
Entrance hole to floor: 7 in. (18 cm)
Floor size: 7 x 7 in. (18 x 18 cm)
Box height: 10 x 10 in. (25 x 25 cm)
Height above ground: 10–17 ft (3–5 m)

Eastern Bluebird, Western Bluebird, Mountain Bluebird

Everyone in North America wants a bluebird to take a nest in one of their birdhouses—the males of all these species have exceptionally bright blue plumage. These birds will use manmade bird boxes placed in fields on fence posts.

Distribution: Eastern Bluebird is resident in the southern part of eastern North America and is a summer visitor as far north as south-eastern Canada; Western Bluebird is resident in the southern Rockies to Pacific coast, and a summer visitor as far north as British Columbia; Mountain Bluebird is resident the southern Rockies, and a summer migrant as far north as Alaska.

Entrance hole: 1½ in. (40 mm)
Entrance hole to floor: 6½ in. (16.5 cm)
Floor size: 5 x 5 in. (12.5 x 12.5 cm)
Box height: 9 in. (23 cm)
Height above ground: 6–8 ft (2–3 m)

European Robin, Spotted Flycatcher

The Robin is a favorite garden bird in Britain and occurs throughout Europe, while the Spotted Flycatcher is a welcome visitor to large gardens with mature trees.

Distribution: Both species breed across Europe and into Asia. Robin is resident in Western Europe, while Spotted Flycatcher is migratory and winters in tropical Africa.

Entrance hole: No entrance hole—box needs to be open-fronted, with a 2½-in. (6-cm) high panel on the lower half of the front of the box.
Floor size: 6 x 6 in. (15 x 15 cm)
Box height: 9 in (23 cm)
Height above ground: 10–17 ft (3–5 m)

House Sparrow, Eurasian Tree Sparrow, Rock Sparrow

These birds will nest in a manmade bird box placed in the corner of a porch eave, although Tree Sparrows in particular sometimes prefer the box to be situated on a tree.

Distribution: House and Tree Sparrows are widespread residents in much of Eurasia, and they have been introduced in other countries, including the United States. Rock Sparrow is a resident in southern Europe.

Entrance hole: 1¼ in. (32 mm) for House and Rock Sparrows; 1⅛ in. (28 mm) for Tree Sparrow
Entrance hole to floor: 6 in. (15 cm)
Floor size: 6 x 6 in. (15 x 15 cm)
Box height: 9 in (23 cm)
Height above ground: 10–17 ft (3–5 m)

Great Spotted Woodpecker, Lesser Spotted Woodpecker, Downy Woodpecker, Hairy Woodpecker

Woodpeckers add a touch of color and excitement to any garden as they clamber about on tree trunks in search of food.

Distribution: Great and Lesser Spotted Woodpeckers are resident in temperate zones across Eurasia, from Britain to the far east of Asia. Downy and Hairy Woodpeckers are resident across North America.

Entrance hole: 2 in. (50 mm) for Great Spotted, 1¾ in. (45 mm) Hairy, 1¼ in. (32 mm) for Lesser Spotted and Downy
Entrance hole to floor: 7 in. (18 cm)
Floor size: 7 x 7 in. (18 x 18 cm) for Great Spotted and Hairy; 5 x 5 in. (12.5 x 12.5 cm) for Lesser Spotted and Hairy
Box height: 10 x 10 in. (25 x 25 cm)
Height above ground: 10–17 ft (3–5 m)

Northern Flicker

A boldly spotted and barred woodpecker, this bird is often seen on the ground, eating ants.

Distribution: Resident in the United States and mainly a summer visitor to Canada; northern populations winter south in Central America.

Entrance hole: 2½ in. (65 mm)
Entrance hole to floor: 19 in. (50 cm)
Floor size: 7 x 7 in. (18 x 18 cm)
Box height: 24 in. (60 cm)
Height above ground: 10–17 ft (3–5 m)

Index

Picture credits

All images are copyright CICO Books except the following:

p. 14: iStock/azndc; p. 20 iStock//Ken Canning; p. 29 iStock//pixtawan; p. 32 iStock Photo; p. 38 iStock Photo; p. 50 iStock//Borislav Filev; p. 56 iStock Photo; p. 62 iStock//Paliman; p. 68 iStock Photo; p. 80, 88 iStock//Andrew_Howe; p. 94 iStock/Prensis; p. 101 iStock/photographereddie; p. 106 iStock/HHelene; p. 113 iStock/Stefan Rotter; p. 116 iStock/Steve_Bramall; p. 119 iStock/Perytskyy; p. 122 iStock/portishead1

Every effort has been made to contact copyright holders and acknowledge sources. Any omissions will be rectified in future printings, if brought to the publisher's attention.

Suppliers

Home-improvement stores and building-supply stores

US

Ace Hardware
www.acehardware.com

The Home Depot
www.homedepot.com

Lowes
www.lowes.com

UK

B&Q
www.diy.com

Homebase
www.homebase.co.uk

Wickes
www.wickes.co.uk

Craft stores

US

Create For Less
www.createforless.com

Michaels
www.michaels.com

Save On Crafts
www.save-on-crafts.com

UK

Craft Superstore
www.craftsuperstore.co.uk

Hobbycraft
www.hobbycraft.co.uk

John Lewis
www.johnlewis.com

Acknowledgments

It is my utmost pleasure to thank the many who have made this book come together. I am very blessed to have such wonderful people in my life. Without you I wouldn't be a whole.

Guido, you were my greatest love, husband, and supporter. Every project I brought for you to view, you gave me such praise and there was always such joy in your face. Just to know that you loved all that I do keeps me wanting to do more. I love you for giving me the freedom to express myself through art.

Ginny, you are my best, best, best friend in the entire world. I love you dearly! Even on those days when I'm crabby, you give me full support and a shoulder to lean on. Thank you from the bottom of my heart for all that you do for me and have done. Thank you for sitting hand in hand with me through each and every page of editing and keeping my wine glass full. I LOVE YOU! My BFF.

I would like to thank CICO Books and all who were involved with the making of the first two beautiful books and the coming of the third book, which combines many projects from the first two. They are stunning! I am truly grateful for that first email and call from Cindy Richards asking if I would be interested in this project.

Sarah, you worked so hard in getting it all put together. I so appreciate everything you've done in the editing and figuring out all my instructions of text and references. It is and always will be a pleasure to work with you.

Steve, the illustrations you've created are amazing and downright awesome. Thank you for all your hard work and know-how. Without you, the book wouldn't be the same.

Alison, you're a fabulous designer—I love how you've made my book look so beautiful!

Penny, thank you for getting the ball rolling and helping me figure out the zip file. I am now a master at it. It was and always will be a pleasure working with you.

James, you are a genius when it comes to photographing my work. The photos are stunning! You are a true artist. Joanna, thank you for designing the set-up for all my creations. The gardens you've chosen are simply filled with beauty and show off each piece beautifully.

Thank you to all the staff in London for all your hard work and to Donna in New York for helping me get all the projects on their way to the UK.

Last but not least, thank you to all who follow me on Facebook and Etsy and love the work I do.

Michele McKee-Orsini